SpringerBriefs in Business

For further volumes:
http://www.springer.com/series/8860

Shunzhong Liu

Innovation Management in Knowledge Intensive Business Services in China

 Springer

Shunzhong Liu
School of Economics
Central China Normal University
Wuhan
People's Republic of China

ISSN 2191-5482   ISSN 2191-5490 (electronic)
ISBN 978-3-642-34675-0 ISBN 978-3-642-34676-7 (eBook)
DOI 10.1007/978-3-642-34676-7
Springer Heidelberg New York Dordrecht London

Library of Congress Control Number: 2012952293

Printed on acid-free paper

Springer is part of Springer Science+Business Media (www.springer.com)

Preface

In the knowledge-based economy, the development of a particular type of services, knowledge intensive business services (KIBS), becomes one of the marking trends in economic evolution. KIBS are private companies or organizations which rely heavily on professional knowledge, i.e., knowledge or expertise related to a specific (technical) discipline or (technical) functional domain to supply intermediate products and services that are knowledge based. The KIBS sector constitutes one of the characteristics of the contemporary economic, and become one of the most dynamic components of the services sector in most industrialized countries. It has become clear that KIBS do innovate and hold an increasingly dynamic and pivotal role in innovation system, and a large share of innovative efforts in KIBS are related to the development of new services.

With the increasing customer expectations, competition and speed of technological development, service firms must constantly look for new approaches to service design and delivery. The management of new service development (NSD) has not only become an important competitive concern in many service industries, but also raised academic interest of researchers in innovation management, marketing management, and operation management. Current research has been a lot of focus on service innovation in developed countries, but very little discussion or thought on that in developing countries.

As gradual opening of the services sector as it has committed to World Trade Organization (WTO), China has paid more and more attention to the service sector in recent years. However, Chinese KIBS are still in its initial stage of development, and the accession to WTO makes them face fiery competition environments. As constant adaptation to a turbulent environment requires a continuous flow of new offers, the management of service innovation has become more important to Chinese KIBS than that of the developed countries.

On the base of empirical research, this book aims to contribute to a better appreciation and understanding of the innovative characteristics of KIBS in China. Data for this study were gathered through enterprise questionnaire investigation in Chinese knowledge intensive business services across four sectors according to Miles's industry classification: management consulting, engineering consultancy, advertising, and software service. This book is organized in the following ways. Chapter 1 introduces the Concepts and Characteristics of Knowledge Intensive Business Services. Chapter 2 describes Innovative Characteristics of Knowledge Intensive Business Services in China. Chapter 3 explores New Service Development Performance within knowledge intensive business services context. Chapter 4 discusses Determinants of Top Performing New Service Development Activities. Chapter 5 explores Knowledge Intensive Service Activities in Chinese Software Industry.

The primary market for this book is faculty in innovation and operation management and their graduate and undergraduate students who have an interest in service innovation. Business students who have an entrepreneurial desire to start their own knowledge intensive business service firms will find the book of interest. Practitioners who are responsible for the marketing and innovation also will find this a readable book that contains useful ideas.

Contents

Chapter 1
The Concepts and Characteristics of Knowledge Intensive Business Services

In the knowledge-based economy, the development of a particular type of services, knowledge intensive business services (KIBS), become one of the marking trends in economic evolution. The KIBS sector constitutes one of the characteristics of the contemporary economic (Muller and Zenker 2001), and become one of the most dynamic components of the services sector in most industrialized countries (Strambach 2001). According to the findings of the innovation survey on the Dutch service industry (Brouwer and Kleinecht 1995), service suppliers are commonly innovating. It has become clear that KIBS do innovate and hold an increasingly dynamic and pivotal role in innovation system(Gallouj and Weinstein 1997), and a large share of innovative efforts in KIBS are related to the development of new services (Den Hertog 2000; Muller and Zenker 2001).

Service innovation is an offering not previously available to the firm's customers that results from either an addition to the current mix of services or from changes made to the service delivery process (Menor and Roth 2007). With the increasing customer expectations, competition and speed of technological development, service firms must constantly look for new approaches to service design and delivery (Smith et al. 2007). The management of new service development (NSD) has not only become an important competitive concern in many service industries (Menor et al. 2002), but also raised academic interest of researchers in innovation management, marketing management and operation management (de Brentani 1989; Thwaites 1992; Cooper et al. 1994; Johne and Story 1998; Story and Kelly 2001; Menor and Roth 2007).

Since the 1990s, there has been a recent upsurge in academic interest related to innovation in KIBS in developed countries (Miles et al. 1995; Howells and Roberts 2000; Muller and Zenker 2001; Tether and Hipp 2002, Hipp and Grupp 2005), but very little discussion or thought on that in developing countries. Providing services encompassing a high intellectual value-added for other firms, KIBS have constituted one important characteristic of the rise of the knowledge-based economy (Muller and Zenker 2001). However, Chinese KIBS are still in their initial stage of

S. Liu, *Innovation Management in Knowledge Intensive Business Services in China*, SpringerBriefs in Business, DOI: 10.1007/978-3-642-34676-7_1,

development, and the accession to World Trade Organization (WTO) makes them face fiery competition environments (Liu 2009). As constant adaptation to a turbulent environment requires a continuous flow of new offers (Stevens and Dimitriadis 2005), the management of NSD has become more important to Chinese service firms than that of the developed countries (Liu 2009). On the base of empirical research, this book aims to contribute to a better appreciation and understanding of the innovative characteristics of KIBS in China.

1.1 Concept of Knowledge Intensive Business Services

Knowledge intensive business services cover a rather wide range of services. On the base of different purpose of research, researchers have given a diverse definition of KIBS.

Miles et al. (1995) argue that KIBS are activities providing services for other businesses with the intention to result in the creation, accumulation or dissemination of knowledge. They define KIBS as.

- private companies or organizations;
- relying heavily on professional knowledge, i.e. knowledge or expertise related to a specific (technical) discipline or (technical) functional domain; and,
- supplying intermediate products and services that are knowledge based.

Windrum and Tomlinson (1999) argue that co-production of new knowledge and material artifacts with their clients are the product that clients wish to purchase. They define KIBS as "private sector organizations that rely on professional knowledge or expertise relating to a specific technical or functional domain". They believe that KIBS are primary sources of information and knowledge or else their services form key intermediate inputs in the products or production processes of other businesses.

Muller and Zenker (2001) argue that KIBS hold a specific position in innovation systems because they play a two-fold role: Firstly, they act as external knowledge source and contribute to innovations in their client firms and secondly, KIBS introduce internal innovations, provide mostly highly-qualified workplaces and contribute to economic performance and growth. According to the role of KIBS in innovation systems, they define KIBS as "firms performing, mainly for other firms, services encompassing a high intellectual value-added".

It is argued that KIBS function either as a facilitator, carrier or source of innovation (den Hertog 2000), we adopt Miles et al. (1995)'s definition in this book.

Knowledge intensive business services are private companies or organizations which rely heavily on professional knowledge, i.e. knowledge or expertise related to a specific (technical) discipline or (technical) functional domain, to supply intermediate products and services that are knowledge based.

1.2 Typology of Knowledge Intensive Business Services

Knowledge-intensive business services form a category of service activities which are often highly innovative in their own right, as well as facilitating innovation in other economic sectors, including both industrial and manufacturing sectors (Den Hertog 2000). On the base of different purposes of research, researchers have mapped a diverse typology of KIBS (Lee et al. 2003). For instance, Miles et al. (1995) divide KIBS into two categories: KIBS I and KIBS II. KIBS I are traditional professional services and liable to be intensive users of new technology, which include marketing/advertising, training (other than in new technologies), design (other than that involving new technologies), financial services (e.g. securities and stock-market-related activities), office services (other than those involving new office equipment, and excluding "physical" services like cleaning), building services (e.g. architecture; surveying; construction engineering, but excluding services involving new IT equipment such as building energy management systems), management consultancy (other than that involving new technology), accounting and bookkeeping, legal services and environmental services (not involving new technology, e.g. environmental law; and not based on old technology e.g. elementary waste disposal services). KIBS II are related to emerging technologies and technological challenges, which include computer networks/telematics (e.g. VANs, on-line databases), some telecommunications (especially new business services), software, computer-related services (e.g. facilities management), training in new technologies, design involving new technologies, office services involving new office equipment), building services (centrally involving new IT equipment such a building energy management systems), management consultancy involving new technology, technical engineering, environmental services involving new technology (e.g. remediation; monitoring), scientific/laboratory services, R&D consultancy and high-tech boutiques.

Lee et al. (2003) argue that KIBS includes public, private, and hybrid suppliers that provide communications management services, R&D services, management consulting services, IT consulting services, employment agency services, engineering consultancy, training services, and contract management services.

Windrum and Tomlinson (1999) argue that KIBS firms are primary sources of information and knowledge or else their services form key intermediate inputs in the products or production processes of other businesses. They include accounting and book-keeping services, architecture, surveying and other construction services, banking and other financial services, computer and IT-related services, design services, environmental services, facility management services, insurance services, legal services, management consultancy, market research, marketing and advertising services, press and news agencies, R&D consultancy services, real estate, telecommunication services, technical engineering services, technology-related training, labour recruitment and provision of technical personnel as KIBS.

Following Miles et al. (1995) KIBS classification, KIBS can be divided into two categories: KIBS I and KIBS II. KIBS I are traditional professional services, liable to be intensive users of new technology. KIBS II are related to emerging technologies and technological challenges. In this study, we will focus on typical KIBS in China, such as engineering, IT-services, advertising and management consultancy, and investigate the innovative characteristic of them in Chinese context. Advertising and management consultancy are more traditional professional services (P-KIBS). Engineering and IT-services are users of scientific and technological knowledge (T-KIBS).

1.3 Characteristic of Knowledge Intensive Business Service

Specialised expert knowledge, research and development ability, and problem solving know-how are the real products of knowledge-intensive services. (Strambach 1997). Knowledge-intensive implies that knowledge is the most important factor which can be an input but also an output (Fiocca and Gianola 2003). Fiocca and Gianola (2003) sum up the features of knowledge intensive business service firms as follows.

First, KIBS may not only be considered as information-intensive. Knowledge is a stock of experience, not only a flow of information and data. Knowledge is value added information.

Second, KIBS are composed by skilled people with exceptional expertise.

Third, KIBS distinguish themself for having a specific perceived core competence. A remark: every firm has got its own expertise. If we have to define a knowledge-intensive firm only as a firm with a specific knowledge and competence, every firm would be considered as knowledge-intensive. The difference is that a KIBS firm makes its knowledge as its "core product" and source of competitive advantage. Don't forget that knowledge and competence are strictly related to people.

Forth, KIBS are strictly in touch with their environment. They change according to their environment: as consequence environment and society reflect, "use", influence these firms. They grow together.

Finally, in KIBS, the specific knowledge is not only "embedded" in people who work in, but also in firms' routine and culture. People convert their knowledge into physical forms when they write books, design buildings, create computer software. Conversely, people may gain knowledge by reading books, studying buildings, running computer programs. People also translate their knowledge into firm's routine, job descriptions, strategies and culture.

Knowledge intensive business services are characterized by their ability to collect information and knowledge externally and to transform these in combination with internal knowledge into service outputs, which are often highly customized to particular user's requirements (Tether and Hipp 2002). According to Tether and Hipp (2002), the characteristic of knowledge intensive business service includes the close interaction between production and consumption, the intangible nature of service outputs, the key role of human resources in service provision, the critical role of organizational factors in firms' performance, and the weakness of intellectual property protection in services.

A large majority of service suppliers do not perform R&D in the traditional sense (R&D department), but a large portion of R&D activities that take place in service suppliers are performed by the other departments (Lee et al. 2003). They often, however, place more emphasis on other innovation-related activities, for example training, and on organizational innovation (Howells 2003.) Miles et al. (1995) suggest that KIBS innovations involve several typical features as follows.

First, lead times seem typically to be long ones. However, in some cases, after having learned the trick, it is possible to speed up the innovation process quite considerably, especially in those cases in which the production of a tangible product is included.

Second, interactions between supply and demand are important in innovation processes. Co-development and interaction with clients are extremely important in developing new services.

Third, process innovations, related to new organizational structures and interaction patterns, are particularly important in the KIBS.

Forth, new work patterns and routines, which require substantial organizational change, are often associated with the KIBS innovations.

Fifth, appropriability and intellectual property problems are experienced very unevenly across our cases. Services embodied in tangible goods and a delivery system can prevent easy imitation and increasing the opportunity to appropriate the value of innovation. However, the more process-oriented and organizational types of innovations are both more difficult to protect.

Finally, standardization is needed for a service innovation to develop.

References

Brouwer E, Kleinecht A (1995) An innovation survey in services: the experience with the cis questionnaire in the Netherlands. STI Review 16:141

Cooper RG, Easingwood CJ, Edgett S, Kleinschmidt EJ, Storey C (1994) What distinguishes the top performing new products in financial services? J Prod Innov Manag 11(4):281–299

de Brentani U (1989) Success and failure in new industrial services. J Prod Innov Manag 6(4):239–258

Den Hertog P (2000) Knowledge-intensive business services as co-producers of innovation. Int J Innov Manag 4(4):491–528

Fiocca R, Gianola A (2003) Network analysis of knowledge-intensive services, IMP conference 2003, Lugano (CH), 4–6 Sept 2003, Work-in-progress pape

Gallouj F, Weinstein O (1997) Innovation in services. Res Policy 26(4/5):537–556

Hipp C, Grupp H (2005) Innovation in the service sector: the demand for service-specific innovation measurement concepts and typologies. Res Policy 34(4):517–535

Howells J (2003) Barriers to innovation and technology transfer in services. Tech Monitor, May-Jun, pp 29–35

Howells J, Roberts J (2000) From innovation systems to knowledge systems. Prometheus, 18(1):17–31

Johne A, Storey C (1998) New service development: a review of the literature and annotated bibliography. Eur J Mark 32(3/4):184–251

Lee K, Shim S, Jeong B, Hwang J (2003) Knowledge intensive service activities (KISA) in Korea's innovation system. OECD, Report 2

Liu S (2009) Organizational culture and new service development performance: insights from knowledge intensive business service. Int J Innov Manag 13(3):371–392

Menor LJ, Roth AV (2007) New service development competence in retail banking: construct development and measurement validation. J Oper Manag 25(4):825–846

Menor LJ, Tatikonda MV, Sampson SE (2002) New service development: areas for exploitation and exploration. J Oper Manag 20(2):135–157

Miles I, Kastrinos N, Flanagan K, Bilderbeek R, Hertog P, Huntink W, Bouman M (1995) Knowledge intensive business services: users, carriers and sources of innovation, Rappon pour DG13 SPRINT-EIMS

Muller E, Zenker A (2001) Business services as actors of knowledge transformation: the role of KIBS in regional and national innovation systems. Res Policy 30(9):1501–1516

Smith AM, Fischbacher M, Wilson FA (2007) New service development: from panoramas to precision. Europ Manag J 25(5):370–383

Stevens E, Dimitriadis S (2005) Managing the new service development process: towards a systemic model. Eur J Mark 39(1/2):175–198

Storey C, Kelly D (2001) Measuring the performance of new service development activities. Serv Ind J 21(2):71–90

Strambach S (1997) Knowledge-intensive services and innovation in Germany. Report for TSER project, University of Stuttgart

Strambach S (2001) Innovation process and the role of knowledge-intensive business services. In: Kulicke KM, Zenker A (eds) Innovation networks—concepts and challenges in the European perspective, Physica-Verlag, Heidelberg, New York, pp 53–68

Tether BS, Hipp C (2002) Knowledge intensive, technical and other services: patterns of competitiveness and innovation compared. Technol Anal Strateg Manag 14(2):163–182

Thwaites D (1992) Organizational influences on the new product development process in financial services. J Prod Innov Manag 9(4):303–313

Windrum P, Tomlinson M (1999) Knowledge intensive services and international competitiveness: a four country comparison. Technol Anal Strateg Manag 11(3):391–408

Chapter 2
Innovative Characteristics of Knowledge Intensive Business Services

2.1 Theory of Service Innovation

2.1.1 The Characteristic of Service Innovation

The service concept represents the operational blueprint that communicates to customers and employees what they should expect to receive and to give (Fitzsimmons and Fitzsimmons 2001). The transformation of any service offering—what the customer receives—either incremental or radical, will require the transformation of some elements of the service concept (Stevens and Dimitriadis 2004). A study of the new modes of innovation in services (Gadrey et al. 1995) has shown that producing a service is to organize a solution to a problem (a treatment, an operation) which does not principally involve supplying a good. In producing a service, the producer is to place a bundle of capabilities and competences (human, technological and organizational) at the disposal of a client and to organize a solution, which may be given to varying degrees of precision (Gadrey et al. 1995). It suggests that apart from technological capabilities, human and organizational capabilities are also important for providing new services.

KIBS are characterized by their abilities to collect information and knowledge externally and to transform these in combination with internal knowledge into service outputs, which are often highly customized to particular user's requirements, so close customer relations often play a decisive role in the provision of these services (Tether and Hipp 2002). Hipp et al. (2000) suggest that knowledge intensive business service suppliers are likely to produce specialized service for specific clients. The customized services, such as consulting and advisory services, often based on more tacit forms of knowledge, services often emerge as a result of co-production between the actual service provider and its client (Den Hertog 2000). The close interaction between production and consumption is one of the innovative characteristics of KIBS, this interaction can be so close that the service cannot be provided without both the service user and provider taking part in its

S. Liu, *Innovation Management in Knowledge Intensive Business Services in China*, SpringerBriefs in Business, DOI: 10.1007/978-3-642-34676-7_2,
© The Author(s) 2013

provision (Tether and Hipp 2002). A study of the characteristics of service (Tether and Hipp 2002) has shown that service firms tend to concentrate on quality and flexibility rather than on price (although some do compete on price), and that, particularly amongst knowledge intensive and technical service firms, this is reflected in the large proportion of income earned from providing services adapted to individual client's needs. It suggests that the most services provided by KIBS are none standardized services, and close interaction with customers is very important in the course of providing new services.

Cohen and Levinthal (1990) point out that a firm's absorptive capacity—its ability to assimilate new information—is closely related to its organizational routines, and the diversity (i.e., the level and distribution) of expertise within an organization. In addition to complementary knowledge and technical competencies, the quality of interaction depends on the practices, beliefs, values, routines and culture of the client organization. If the use of these knowledge intensive business services is to be encouraged then trust must be built up across the client organization (Windrum and Tomlinson 1999). The researches above suggest that the interface between the service provider and its clients must be re-designed in the process of service innovation.

Using software industry as example, Broch and Isaksen (2004) find that KIBS require close contact with clients in order to be innovative and build innovative capacity, and 50 % of the software firms work extensively at clients' offices, and most software firms have regular face-to-face meetings with their clients, and after a sale most software firms continue to collaborate with clients by phone, e-mail etc. This suggests that the contact with clients is very most important factor in the service innovation in KIBS. According to the degree of contact with customers, the contact manners can be classified into three categories as follows: working at clients' office for a long time, regular face-to-face meetings with clients and contacting with clients after providing service. Working at clients' office for a long time is the strongest contact manner; however, contacting with clients after providing service is the weakest one.

Once a kind of service innovation takes place, it may soon cause organizational change so technological innovation and organizational innovation should be undertaken together (Lee et al. 2003). Typically, there is a close relation between the technologies employed and the organizational form of the service, which also has implications for the process of service provision and the nature of the services provided, and hangs to any one of these frequently require or bring about change to the others (Tether and Hipp 2002). Knowledge intensive business service suppliers are lead users of information and communication technology (Miles and Boden 2000), integration of ICT into many knowledge intensive services has led to a new paradigm of service innovation.

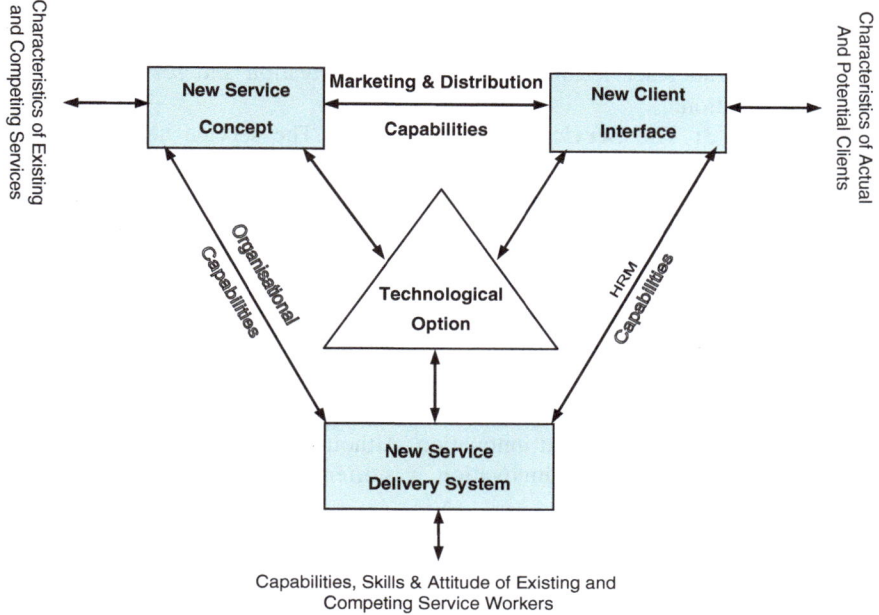

Fig. 2.1 A four-dimensional model of service innovation[1]

2.1.2 Model of Service Innovation

In order to discuss, map and analyze the diversity of innovations in greater details and in a structured way, Den Hertog (2000) introduce a four-dimensional model of service innovation as shown in Fig. 2.1. Beside technology innovation, the model points to the significance of such non-technological factors in innovation such as new service concepts, client interfaces and service delivery system.

According to the model of service innovation, any of the service innovation involves some combination of the dimensions as follows.

Dimension 1: The Service Concept: Although not all service innovations have a strong novel conceptual element, many service innovation involve more intangible characteristics, such as new ideas of how to organize a solution to a problem. Conceptual innovations are much more likely to be found in service firms than in pure manufacturing firms.

Dimension 2: The Client Interface: The client interface innovation is the design of the interface between the service provider and the clients, these interfaces are the focus of a good deal of service innovations. In business services in particular, clients are regularly part and parcel of the production of the service product. The interactions

---

[1] Den Hertog 2000

between the service provider and the clients can be a source of innovation. According to the high degree of co-innovation in new service development, the client interface innovation includes service providing manner innovation and service incepting manner innovation.

Dimension 3: The Service Delivery System: The service delivery system innovation refers to the internal organizational arrangements that have to be managed to allow service workers to perform their job properly, to develop and offer innovative services. It is closely related to the question of how to empower and facilitate employee, therefore, they can perform their jobs and deliver service products adequately.

Dimension 4: Technological Options: Service innovation is possible without technological innovation, but there is a wide range of relationships between technology innovation and non-technological innovation in practice. Technology mainly plays a role as a facilitating or enabling factor, something much closer to supply-push, technology-driven innovation. Although IT is certainly not the only relevant technology in service innovation, it is often perceived as the great enabler of service innovation.

In practice, it may be the combination of the four dimensions that ultimately characters each particular service innovation. Because any service innovation involves some combination of the above-mentioned dimensions of service innovation, which leads us to our two hypotheses as follows.

H1: the T-KIBS and P-KIBS have the same characteristics in the four service innovation dimensions.

H2: the importance of four service innovation dimensions is the same in the service innovation of KIBS.

2.2 Method

2.2.1 Sample

Data for this study were gathered through enterprise questionnaire investigation. We followed a four-step procedure. Firstly, through an exhaustive search on internet, we got the initial sample of knowledge intensive business service firms in Wuhan, the People's Republic of China. Secondly, we listed all the office building where the sample firms located, and then classify these office buildings into five groups according to their location. Thirdly, we used a pretesting of the questionnaire for clarity and relevance through face-to-face interviews with manager in the six firms. Finally, each of the five investigators survey manager, who would like to finish the pre-tested questionnaires, of the knowledge intensive business service firms in one group of the office buildings through face-to-face interviews.

We have surveyed 102 firms in which firms with no more than five employees were not included. The number of respondents for T-KIBS and P-KIBS were 46 (45 %) and 56 (55 %) respectively.

2.2.2 Measures

According to the model of service innovation (Den Hertog 2000), variables measuring service innovation included concept innovation(1 = my company has provided service with new concept to customers; 0 = my company has not provided services with new concept to customers), providing interface innovation(1 = my company has provided service with new manners; 0 = my company has not provided service with new manners), incepting interface innovation(1 = clients have accepted service with new manners; 0 = clients have not accepted service with new manners), organizational innovation (1 = my company has changed organizational structure and personnel to improve the efficiency for providing service; 0 = my company has not changed organizational structure and personnel to improve the efficiency for providing service), IT technology innovation(1 = my company has innovated in IT technology to improve the efficiency for providing service; 0 = my company has not innovated in IT technology to improve the efficiency for providing service), and the other technology innovation (1 = my company has innovated in the other technology to improve the efficiency for providing service; 0 = my company has not innovated in the other technology to improve the efficiency for providing service).

We used a variable to measure the degree of service customization, and asked them to rank on a scale of 1 (fully standardized service) to 5 (fully customized service).

The survey also asked the firm what is the most important contacting manner that the firm contacts with clients. We used two-point scale to measure if the employees work at clients' office for a long time, or employees in KIBS has regular face-to-face meetings with clients, or contact with clients after providing service.

2.3 Analysis and Results

2.3.1 Descriptive Statistics

KIBS need close contact with clients in order to be innovative and build innovative capacity. The Fig. 2.2 reveals that 68 firms surveyed continue to collaborate with clients after providing services, 21 firms work extensively at clients' offices, and 13 firms have regular face-to-face meetings with their clients. KIBS very often negotiate new contracts or develop new solutions based on signals from clients, thus, many activities should take place before providing services.

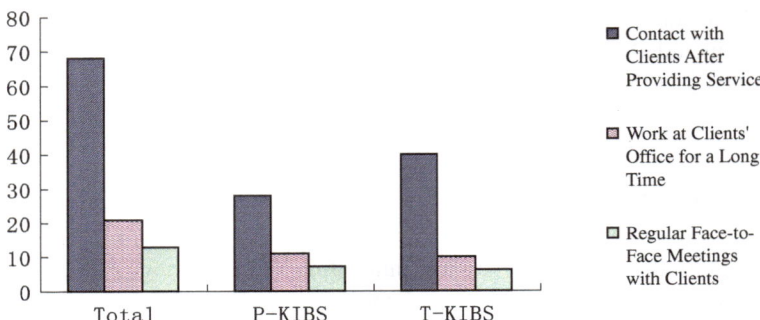

Fig. 2.2 The contacting manner of providing service in T-KIBS and P-KIBS

Though contact with clients after providing service is the weakest contact manner, it is the most important contact manner for KIBS in China. Working at clients' office for a long time and regular face-to-face meetings with clients are less important contact manner than contacting with clients after providing service, though they are stronger contact manner. It suggests that KIBS in China weakly contact with clients in providing service to clients.

We computed logical inclusive of two variables about service interface innovation (service providing manner and service incepting manner), and used this value to measure whether or not the firm has service interface innovation. As the same, we counted whether or not the firm has technology innovation. The number of firms which has provided each dimension of new services to customer is shown in the table below.

As shown in Table 2.1, the number of firms surveyed which have provided the service concept innovation is almost the same as that of organization innovation and technology innovation. Approximately 90 % of the firms surveyed have provided the three dimensions of service innovation above in recent 3 years. Only 74.51 % of the firms surveyed have provided interface innovation in recent 3 years. For the service interface innovation, 64.71 % of the firms surveyed have provided service with new manners, and 64.71 % of the firms surveyed have considered that the clients have accepted services with new manners. For technology innovation, 84.31 % of the firms surveyed have provided the IT technology innovation in recent 3 years, and only 50.98 % of the firms surveyed have provided the other technology innovation in recent 3 years. The data above suggest that the capacity of interface innovation of firms surveyed is lower than that of the other dimensions of service innovation, and also indicate that integration of IT into many knowledge intensive services has also led to a new paradigm of service innovation in China.

Table 2.1 The characteristic of service innovation

Dimension	Total		P-KIBS		T-KIBS	
	N	%	N	%	N	%
Concept innovation	92	90.2	42	91.3	50	89.29
Interface innovation	76	74.51	34	73.91	42	75
Providing manner	66	64.71	32	69.57	34	60.71
Incepting manner	66	64.71	28	60.87	38	67.86
Organization innovation	91	89.22	42	91.3	49	87.5
Technology innovation	94	92.16	44	95.65	50	89.29
IT	86	84.31	40	86.96	46	82.14
The others technology	52	50.98	25	54.35	27	48.21

Table 2.2 The degree of service customization

Categories	N	Mean	Std. deviation
P-KIBS	46	1.83	0.64
T-KIBS	56	1.79	0.62
Total	102	1.80	0.63
Mann–Whitney U = 1248, Z = −0.304, Asymp. Sig. (2-tailed) = 0.761			

2.3.2 Statistical Analysis

Using Mann–Whitney U test, we examined whether or not the means of the degree of service customization differ between P-KIBS and T-KIBS.

As shown in Table 2.2, the two-sided asymptotic significance of the Mann–Whitney U statistics is greater than 0.10, so it's safe to say that the differences are due to chance variation, which implies that there are no differences between P-KIBS and T-KIBS in the degree of service customization. The data in Table 2.2 also show that the services provided by KIBS in China appear a very high degree of standardization.

Using χ^2-test, we examined the association of these four service innovation dimensions (service concept innovation, service interface innovation, organizational innovation, and technology innovation) with respect to the categories of KIBS. The result is shown in Table 2.3.

As shown in Table 2.3, all the two-sided asymptotic significance of the Chi square statistics is greater than 0.10, so it's safe to say that the differences are due to chance variation, which implies that the results above support the hypothesis 1. The P-KIBS and T-KIBS have the same characteristic in the four service innovation dimensions in China.

Because all the variables for measuring service innovation are a two-point scale and the variables are measured on the same company, we use Cochran's Q-test to test hypothesis that the important of four service innovation dimensions is the same in service innovation in KIBS.

Table 2.3 Association of innovation dimensions with categories of KIBS

Dimension	Pearson chi square	df	Asymp. sig.
Concept innovation	0.116	1	0.733
Interface innovation	0.016	1	0.9
Organization innovation	0.38	1	0.538
Technology innovation	1.416	1	0.234

Table 2.4 Relationship of the four service innovation dimensions

	Total	Without concept innovation	Without interface innovation	Without organization innovation	Without technology innovation
N	102	102	102	102	102
Cochran's Q	102.000	18.600	0.583	16.222	13.771
df	3	2	2	2	2
Asymp. Sig.	0.000	0.000	0.747	0.000	0.001

As shown in Table 2.4, for the four dimensions of service innovation, the two-sided asymptotic significance of the Cochran's Q statistics is less than 0.05, it's safe to say that the differences are not due to chance variation, which implies that there are significant difference among the four dimension of service innovation. The result rejects the hypothesis 2 that the four service innovation dimensions are the same important in service innovation in KIBS. For the four dimensions of service except concept innovation, the two-sided asymptotic significance of the Cochran's Q statistics is less than 0.05, it's safe to say that the differences are not due to chance variation, which implies that there are significant difference among the interface innovation, organization innovation and technology innovation. For the four dimensions except organization innovation or technology innovation, the result is the same as that except concept innovation. However, for the four dimensions of service except interface innovation, the two-sided asymptotic significance of the Cochran's Q statistics is greater than 0.05, it's safe to say that the differences are due to chance variation, which implies that there are no significant difference among the concept innovation, organization innovation and technology innovation.

The result above suggests that the four service innovation dimensions have different role in the service innovation of KIBS in China, and that the four service innovation except interface innovation are the same important in service innovation in KIBS and the interface innovation are less important than that of the other three service innovation.

2.4 Discussion

The purpose of this study is to analyze characteristics of service innovation of KIBS in China. The research results specifically indicate that the services provided by Chinese KIBS stand on a very high degree of standardization and KIBS weakly contact clients during service innovation, the P-KIBS and T-KIBS have the same characteristics in the four service innovation dimensions, and the capacity of interface innovation is weaker than that of the other three service innovation for Chinese KIBS.

According to Antonelli (1999), KIBS firms perform two important functions in economic system: firstly, as containers of proprietary 'quasi-generic' knowledge, extracted by means of repeated interactions with customers and scientific community; secondly, as an interface between that knowledge and the tactic knowledge buried in routines of firms. The knowledge intensive service suppliers which provide specialized services are more likely to undertake innovations than to standardized service providers, so the specialized suppliers tend more to suit specific users than standardized suppliers (lee et al. 2003). Because high degree of customization in the output of KIBS, innovation activities of which are relatively more oriented towards product innovation, while new service can not be provide without the service user and provider taking part in its provision (Tether and Hipp 2002). Larsen (2000) distinguishes three types of knowledge that user firms gain from KIS suppliers: core, operational, and peripheral knowledge. In providing core knowledge, the interaction between supplier and user is stronger than that of in providing operational and peripheral knowledge. As the results shown, the most service clients gotten are standardized and the contact of provider and user are weak in providing service.

In order to improve innovation capacity of KIBS in China, firms should pay attention to the contacting manners and the degree of standardization in providing the knowledge intensive services to client, some things should be carried out for KIBS. Firstly, the employee of service supplier should work at clients' office for a long time and have regular face-to-face meetings with clients in order to have better interactions with users and understanding of users' features and requirements. So the customized services which build users' features and requirements into their services to benefit the customers can be produced and provided to client. Secondly, in contacting with client, the firm should have better understand the structure and characteristic of clients' organization, and design the interface between the service provider and its clients to make the way, through which the service provider interacts with the client to suit the service provided, thus the customized service can be more easily delivered to client. Finally, the firms should use stronger contact manner in providing service to understand client's needs better. Through the close interaction with client, the KIBS can develop its competencies to provide core knowledge to clients.

References

Antonelli C (1999) The microdynamics of technological change. Routledge, London
Broch M, Isaksen A (2004) Knowledge intensive service activities and innovation in the Norwegian software industry, STEP REPORT, 03

Cohen WM, Levinthal DA (1990) Absorptive capacity: a new perspective on learning and innovation. Adm Sci Q 35(1):128–152

Den Hertog P (2000) Knowledge-intensive business services as co-producers of innovation. Int J Innov Manag 4(4):491–528

Fitzsimmons JA, Fitzsimmons MJ (2001) Service management, 3rd edn. McGraw-Hill, New York

Gadrey J, Gallouj F, Weinstein O (1995) New modes of innovation. How services benefit industry. Int J Serv Ind Manag 6(3):4–16

Hertog P (2000) Knowledge-intensive business services as co-producers of innovation. Int J Innov Manag 4(4):495

Hipp C, Tether B, Miles I (2000) The incidence and effects of innovation in services: evidence from Germany. Int J Innov Manag 4(4):417–454

Larsen JN (2000) Supplier-user Interaction in Knowledge? intensive Business Services: Types of Expertise and Modes of Organization. In: Miles I, Boden M (eds) Services and the Knowledge-Based Economy. Routledge, London and New York, pp 146–154

Lee K, Shim S, Jeong B, Hwang J (2003) Knowledge intensive service activities (KISA) in Korea's Innovation System, OECD Report, 2

Miles I, Boden M (2000) Introduction: are services special? In: Miles I, Boden M (eds) Services and the knowledge based economy. Continuum, London

Stevens E, Dimitriadis S (2004) New service development through the lens of organisational learning: evidence from longitudinal case studies. J Bus Res 57(10):1074–1084

Tether BS, Hipp C (2002) Knowledge intensive, technical and other services: patterns of competitiveness and innovation compared. Technol Anal Strateg Manag 14(2):163–182

Windrum P, Tomlinson M (1999) Knowledge intensive services and international competitiveness: a four country comparison. Technol Anal Strateg Manag 11(3):391–408

Chapter 3
New Service Development Performance

3.1 Introduction

It is widely accepted that you cannot manage what you can not measure (Scholey 2005). Performance measurement, which plays a key role in translating an organization's strategy into desired behaviors and results (Van der Stede et al. 2006), is the process of quantifying past action (Neely 1998). The need of performance measurement systems at different levels of decision-making, either in the manufacturing or service contexts, is undoubtedly not something new (Bititci et al. 2005). Griffin and Page (1993) believe that "If underlying dimensions could be identified, then regardless of which specific measures were used to quantify the firm's performance in a particular dimension, we might ultimately be able to help firms determine whether they were missing any aspects of measurement that would help provide them with a more balanced view of their performance".

The importance of performance measurement is generally recognized in the literature and by industry (Driva et al. 2001; Alegre et al. 2006). In achieving long-term NSD success, it is vital that a firm have effective mechanisms for assessing NSD success or failure (Storey and Kelly 2001). In today's service-oriented and knowledge-based economies, organizational leaders are quickly realizing that their organizations can no longer compete on past success factors such as assets, products, or pricing (Pangarkar and Kirkwood 2008). In order to survive and stay competitive in the increasingly dynamic environment, organizations should be sensitive to emergent changes, encourage the usage of knowledge management and foster learning (Blazevica and Lievens 2004). To response to criticisms of traditional forms of accounting reports for knowledge-based firms (Bose and Thomas 2007), Kaplan and Norton (1992, 1996) have proposed the balanced scorecard (BSC) as a means to evaluate corporate performance from four related perspectives: the financial perspective; the customer perspective; the internal-business-process perspective; and the learning and growth perspective. The BSC concept suggests that NSD performance can be best accessed by taking a balanced view across these four dimensions.

S. Liu, *Innovation Management in Knowledge Intensive Business Services in China*, 17
SpringerBriefs in Business, DOI: 10.1007/978-3-642-34676-7_3,
© The Author(s) 2013

Since the 1990s, there has been a recent upsurge in academic interest related to NSD performance measures in the developed countries (de Brentani 1989; Cooper et al. 1994; Storey and Kelly 2001; Voss et al. 1992; Ottenbacher 2006; Ahn et al. 2006; Menor and Roth 2007). While each paper is interesting, there are some problems in the current NSD performance measures. First, though this research has tended to use a variety of different measures of success which can be classified into subset of four perspectives with BSC framework, none of the measures include all the four performance dimensions. In particular, with few exceptions (Ahn et al. 2006), learning and growth performance have been the topic of very little research on NSD performance measure. To date, how to integrate the learning and growth dimension is not addressed in depth. Managers have an incentive to concentrate on those activities for which their performance is measured, often at the expense of other relevant but non-measured activities (Hopwood 1974). A number of authors have argued that broadening the set of performance measures enhances organizational performance (e.g., Lingle and Schiemannm 1996; Van der Stede et al. 2006) and reduces such dysfunction effects (Lillis 2002). Consequently, the development of a diverse set of NSD performance measures will redound to driving all areas important to NSD.

Second, the research has resulted in an impressive amount of literature; however, there is very little consensus amongst their studies regarding how best to measure NSD performance: which dimensions of success to include and how to set about measuring these dimensions. According to research on dimension of success in new product development (Hart 1993), the way in which NSD success is defined may influences the findings which describe the factors contributing to NSD success. As a result, different success measures make it difficult to draw generalizations across the studies (Griffin and Page 1993). Measuring and empirically testing an NSD performance scale may contribute towards consensus measure NSD performance and make generalizable conclusions more easily drawn.

Finally, the targeted firms are, as a rule, very reluctant to disclose sensitive financial information in survey research (Rosenzweig and Roth 2007; Nakos et al. 1998; Hart 1993). Missing values can obscure the results, reduce the precision of calculated statistics, and complicate the theory required (Hill 1997). Researchers have used subjective measures of performance based on the perception of key executives (McCracken et al. 2001) in response to this constraint because the subjective measures of performance are essentially equivalent to those for objective performance (Wall et al. 2004). Though non-financial measures such as product quality and customer satisfaction are far too subjective and susceptible to manipulation (Andrews 1996), recent work in accounting has begun to focus on the use of subjectivity in performance measurement, evaluation, and incentives (Van der Stede et al. 2006). Good measurement is a prerequisite for good empirical science (Menor and Roth 2007); thereby the lack of psychometrically sound or generally accepted multi-item measurement scales has hindered the theory and understanding of NSD.

With the aim to fill these gaps, this chapter contributes to service management research by providing a set of theoretically and psychometrically sound metrics reflecting NSD performance within the BSC frame. The purpose of this study is threefold. First, provide consensus scale reflecting all areas important to NSD. Second, give researchers the ability to compare findings across different studies in NSD performance. Finally, give practitioners a diagnosing and benchmarking tool for NSD performance of their own organization.

3.2 A Literature Review on NSD Performance Dimension

The researchers have used a variety of different measures to assess whether a NSD has been successful (Cooper and Kleinschmidt 1987; de Brentani 1989; Voss et al. 1992; Griffin and Page 1993; Cooper et al. 1994; Storey and Easingwood 1996; Storey and Kelly 2001; Gray et al. 2002; Blazevic and Lievens 2004; Hipp and Grupp 2005; Ahn et al. 2006). Because simple unidimensional measures of performance are inappropriate in any but the most circumscribed NSD endeavors, empirical researchers have relied on measuring performance on a number of prescribed dimensions (Johne and Story 1998). Cooper and Kleinschmidt (1987) use financial performance, window of opportunity and market impact to measure NSD performance. de Brentani (1989) use sales and market share, competitive, other booster and cost. Voss et al. (1992) use financial dimension, competitiveness dimension and quality dimension to measure the results of NSD. Griffin and Page (1993) identify five categories of performance measures: overall firm benefits, programme level benefits, product level benefits, financial benefits and customer acceptance benefits. Cooper et al. (1994) suggest that NSD performance dimensions included financial, relationship enhancement and market development. Storey and Easingwood (1996) indicate that sales performance, enhanced opportunities and profitability are three underlying performance dimensions. Gray et al. (2002) suggest that NSD performance include business performance, marketing performance and other performance (e.g. ROI and overseas sales). Hipp and Grupp (2005) use the following independent dimensions to measure NSD performance: improvement of the quality of the service product, compliance with environmental standards and safety requirements, company internal improvements and improvement of customer performance or productivity. Ahn et al. (2006) use business performance and knowledge performance to measure new product development performance. Blazevic and Lievens (2004) give a unidimensional scale to measure new service development project learning performance. Though a variety of different measures are used by previous literature, they can be classified into subset of four perspectives with BSC framework (Kaplan and Norton 1992, 1996). On investigating how service firms evaluate their NSD activities, Storey and Kelly (2001) suggested that BSC framework is suitable to measure NSD performance.

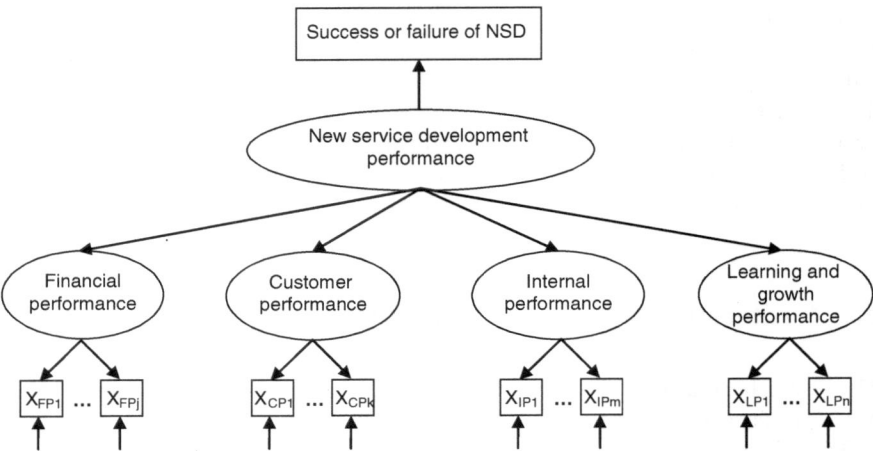

Fig. 3.1 Conceptual model of NSD performance

3.3 Conceptual Framework of NSD Performance

To response to criticisms of traditional forms of accounting reports for knowledge-based firms (Bose and Thomas 2007), Kaplan and Norton (1992, 1996) have proposed the balanced scorecard to evaluate corporate performance from four different perspectives: the financial perspective; the customer perspective; the internal-business-process perspective; and the learning and growth perspective. The four perspectives provide a framework to translate a strategy into operational terms. An important part of BSC framework is the emphasis on the balance and integration between four measurement dimensions. The BSC complements financial measures of past performance with measures of the drivers of future performance, is increasingly being used as multi-dimensional performance assessment systems to help managers achieve their strategic objects (Bhagwat and Sharma 2005). Many firms are implementing a balanced scorecard performance measurement system that tracks measures across the four perspectives (Lipe and Salterio 2000; Bryant et al. 2004). The development of the NSD performance scales within the BSC frame cloud make possible consensus amongst their studies regarding how best to operationalize "success". Our literature review on NSD performance also indicates that a variety of different measures of NSD success can be structured into the BSC framework. Measuring NSD performance within the BSC frame can capture information on all aspects of the NSD activities. Consequently, BSB framework can be use as tool to identify the underlying dimensions of NSD performance and help to achieve a greater homogeneity and comparability among NSD research. Our conceptual model of NSD performance dimensions is depicted in Fig. 3.1.

NSD performance measure is not directly observable, and its study requires scrutiny of the underlying dimensions that reflect such a performance. NSD performance, as that of new product (Griffin and Page 1993), is measures of new service development success and failure. On the base of BSC framework, the extant service management and innovation literature, and a series of interviews conducted with service professionals involved in NSD-related activities, we consider NSD performance to be a latent multidimensional construct inasmuch as its full significance lies beneath the various dimensions that go towards its makeup. These dimensions (i.e. financial performance, customer performance, internal performance, and learning and growth performance) sum up the aspects mentioned previously as the basic elements needed for a firm to focus attention on. According to the BSC framework (Kaplan and Norton 1992), the complexity of NSD requires that managers be able to view performance in the four performance dimensions simultaneously. Thus the four complementary dimensions are the critical dimensions to NSD performance, while this dimension is antecedents to success or failure of NSD. We provide below an operational definition of each NSD performance dimension and review the relevant scholarly literature. The initial set of representative items tapping each construct and its supporting literature are summarized.

3.3.1 Financial Performance

Financial measures sum up the commonly measured economic effects of NSD activities and show whether NSD contribute to the improvement of the firm's economic results. Storey and Kelly (2001) propose that the widely used measure of NSD performance included profitability, shareholder benefits, profit-margin, payback period, costs and development/investment costs. Sales-based measures, which include the level of sales, market share, customer use and sales growth, are often used as financial performance measures (Storey and Kelly 2001). Cooper and Kleinschmidt (1987) use relative profits to sales, profitability level, pay-back period and market share to measure the NPD success and failure from financial perspective. Previous researches suggest that profit, sales, market share and cost are the widely used financial measures to gauge the NSD success or failure (de Brentani 1989; Voss et al. 1992; Griffin and Page 1993; Cooper et al. 1994; Storey and Easingwood 1996; Gray et al. 2002; Ahn et al. 2006). Financial performance measures indicate whether the company's strategy, implementation, and execution are contributing to bottom-line improvement (Kaplan and Norton 1996).

3.3.2 Customer Performance

Customer perspective captures the ability of the organization to provide quality goods and services, the effectiveness of their delivery, and overall customer service and satisfaction (Amaratunga et al. 2001). Storey and Kelly (2001) suggest that customer-based measures of new product performance include customer satisfaction, perceived product quality, customer acquisition and customer retention.

Sampson (1970) defines success from the consumer point of view as a new product which: (i) satisfies new needs, wants or desires; (ii) possesses outstanding performance compared to other products, (iii) benefits from an imaginative combination of product and communication. In the NSD research, the measures, such as superior service, service quality, unique benefits, customer satisfaction and relation enhancement, are widely used to capture the customer performance (de Brentani 1989; Voss et al. 1992; Easingwood and Storey 1993a; Griffin and Page 1993; Storey and Easingwood 1996; Gray et al. 2002; Hipp and Grupp 2005). With the importance of customer orientation readily accepted by managers (Auh and Menguc 2007), performing from its customers' perspective has become a priority for top management (Kaplan and Norton 1992). Customer performance dimension reflects the company's value for its customers, which are the drivers for financial performance.

3.3.3 Internal Process Performance

Internal business processes are the mechanisms through which performance expectations are achieved (Amaratunga et al. 2001). The internal business process measures focus on the internal processes that will have the greatest impact on customer satisfaction and on achieving an organization's financial objectives. Fitzgerald et al. (1991) propose that the performance of the development process itself should be measured on cost, speed and effectiveness. The researcher often use productivity, reliability, delivery capability and better service platform to measure the NSD internal process performance (de Brentani 1989; Voss et al. 1992; Cooper et al. 1994; Storey and Easingwood 1996; Hipp and Grupp 2005). Financial and customer-based measures of performance in NPD (i.e. NSD) represent performance in the marketplace, i.e. they are external to the firm (Storey and Kelly 2001). Managers must focus on critical delivery system to achieve these external performances, so it is important to measure performance internally. The internal-business-process perspective describes the business processes to which the company has to be particularly adapted in order to satisfy its shareholders and customers (Fernandes et al. 2006).

3.3.4 Learning and Growth Performance

The organizational learning and growth perspective involves the changes and improvements which the company needs to realize if it is to make its vision come true (Fernandesa et al. 2006), and identifies the infrastructure that the organization must build to create long-term growth and improvement. Storey and Kelly (2001) suggest that program-level measures should indicate the extent to which the service firm has learnt from the successes and failures of individual projects, and incorporated that learning to develop NSD into an organizational capability offering true competitive advantage in the marketplace. Technical knowledge and market knowledge are two components of knowledge created in the innovation process (Maidique and Zirger 1985; Ahn et al. 2006). Muller and Zenker (2001) believe that employees acquire the knowledge from clients through developing and launching new services. From the perspective of organizational learning, Blazevic and Lievens (2004) believe that organizational learning occurs at different cognitive levels and involves three subprocesses: acquisition, distribution and interpretation. Acquisition relates to the process by which information is obtained. Distribution of information refers to the process of information dissemination between different information sources. Interpretation occurs when information becomes meaningful by sharing perceptions and building cognitive maps. The targets for success keep changing and intense competition requires that organizations make continual improvements to their existing products and processes and have the ability to introduce entirely new processes with expansion capabilities (Kaplan and Norton 1992). Learning and growth perspective provides the infrastructure to enable ambitious objectives in the other three perspectives to be achieved.

3.4 Research Method

On the base of program-level measures, we used knowledge intensive business services (KIBS) as research object to develop a measurement scale for NSD performance. Our scale development and refinement followed Menor and Roth's (2007) structured two-stage multi-item scale-development approach for the selection of appropriate measurement items and the coverage of the construct domain with the desired reliability and validity. In the first stage, an item-to-construct sorting analysis was employed to establish tentative item reliability and validity. In the second stage, we apply confirmatory analyses to derive stronger assessments of the psychometric properties of our multi-item scales reflecting the NSD performance dimensions.

3.4.1 Level of NSD Performance Measure

NSD performance can be measured on a project or overall development process level (Johne and Storey 1998; Voss et al. 1992; Sotry and Kelly 2001), and it is difficult to get some performance measures (i.e. financial measure) for a single project because the different NSD projects often shares the same delivery system (Storey and Kelly 2001). Undue focus on project success (and failure) in the product development literature has resulted in neglect of wider strategic considerations (Johne and Storey 1998). A NSD project might be highly successful in its own right in the short run but do harm within a broader NSD program in the longer run and the reverse might also apply (Johne and Storey 1998). de Brentani (2001) assert that firms undertaking new product development are likely to be involved in, not one, but a portfolio of different types of projects. The determinants of new service development impact not only single projects but also a portfolio of different types of projects. The NSD program refers to the portfolio of NSD projects the service organization has initiated within certain period (Menor and Roth 2007). Johne and Storey (1998) also suggest that "Focusing on (project) failure represents a limited and restricted approach to business development via the organic NSD route because it fails to consider adequately the full span of ways in which NSD can contribute to organic growth". Program-level measures should indicate the extent to which the service firm has learnt from the successes and failures of individual projects, and incorporated that learning to develop NSD into an organizational capability offering true competitive advantage in the market-place (Storey and Kelly 2001). Consequently, in this research, we develop measurement scale for NSD performance on program level rather than individual project level.

3.4.2 Research Subject

We test the proposed measurement scale by focusing on a single industry: knowledge intensive business services. According to World Bank statistics (See http://www.stats.gov.cn), the service sector accounted for 72.2 % of GDP in the developed countries in 2004, whereas the service sector only accounted for 40.7 % in China. With the accession to the WTO, the fulfillment of China's accession commitments will lead to one of the dramatic episodes of liberalization (Mattoo 2003). In 2006, China has newly established 7141 foreign-invested services companies (See http://english.mofcom.gov.cn). To survive this fiery competition environment, the management of NSD has become more important to Chinese service firms than that of the developed countries. Especially, knowledge intensive business service in China is still in its initial stage of development, much lower than that of developed countries, which make them face more fiery competition that that of mass service sectors.

As one of the most dynamic components of the services sector in most industrialized countries (Strambach 2001), KIBS innovate and hold an increasingly dynamic and pivotal role in innovation system (Muller and Zenker 2001; OECD 1997). A large share of innovative efforts in KIBS is related to the development of new services (den Hertog 2000; Muller and Zenker 2001).The research on KIBS should explore relevant organizational characteristics for these sectors. Current research on NSD performance measures has been limited to production-intensive services such as banks, insurances, financial, transport, wholesale and telecommunications (de Brentani 1989; Cooper et al. 1994; Storey and Kelly 2001; Menor and Roth 2007), which result in ignoring learning and growth facts. Kaplan and Norton (1996) stress the need for management to be attentive to the "Learning and growth" component of the BSC and not disregard or minimize its importance to the other perspectives. To date, how to measure the learning and growth performance, except for few researches (Ahn et al. 2006; Blazevic and Lievens 2004), is not addressed in any depth. Specialized expert knowledge, research and development ability, and problem solving know-how are the real products of knowledge intensive services (Strambach 1997); hence KIBS are one of typical object of knowledge-based firms. Consequently, the present study use KIBS as the subject for research into NSD performance.

3.4.3 Item and Scale Construction

Item generation and translation: Content validity is determined by the degree to which questions, tasks, or items on a test are representative of the universe of behavior the test was designed to sample (Gregory 2005). For item generation to provide the basis for claims of content validity, studies should clearly show links between the items and their theoretical domain (Hensley 1999). Multi-item measurement and scale development must be preceded by sound conceptual development of the theoretically important construct(s) being defined (Hinkin 1998; Churchill 1979). The usual methods of ensuring content validity are extensive review of the literature and the preliminary test using expert judges (Stratman and Roth 2002; Jerez-Gómez et al. 2005). In our case, we did an exhaustive overview of the literature and a preliminary test using personal interviews with four managers in KIBS to develop valid constructs. NSD performance were deemed to be content valid at the outset of stage one. The translation process was similar to the parallel translation/double translation method (Avlonitis et al. 2001). A first version of items was prepared in English and then it was translated to Chinese by a panel including professors specialize in service and innovation management and Chinese professional translators. We conducted this study in Chinese language. To be sure that the translation is adequate to be useful as scales in the Western world, we employed two English translators who independently translated the Chinese items back into English. Only minor inconsistencies occurred and resolved through communication with two translators. See Tables 3.1, 3.2, 3.3 and 3.4 for items of each construct.

Purification and pre-testing of measures: For purification and pre-testing of items placed in a common pool, we applied a modified version of Q-Sort method (Nahm et al. 2002). This method accesses reliability and construct validity of questionnaire item that are being prepared for survey research, to systematically refine the construct definitions and item wording. The items and associated constructs were subjected iteratively to four rounds of independent sorting, each with a different panel of practitioners who served as judges (Independent samples of two judges per sorting round). In the first two rounds of the four item-sorting iterations, we employed convenience samples of graduate students and professors specialize in service and innovation management and in the last two, we used practitioner judges with NSD expertise in KIBS. Questionnaire items that were identified as being too ambiguous, as a result of the first stage, are reworded or deleted, in an effort to improve the agreement between the judges. In addition to deleting ambiguous items, some ambiguous items, though fitted within target construct, were reworded according to the judges. Moore and Benbasat (1991) note that if an item is consistently placed by the judges within its intended construct, it is considered to demonstrate convergent validity with that construct and discriminant validity with the others. The degree of agreement between judges within each round of the Q-sorting exercise formed the basis for the statistical assessment of tentative reliability and construct validity (Rosenzweig and Roth 2007). In this study, we use the three evaluation indices to measure inter-judge agreement level: Cohen's k (Cohen 1960), Perreault and Leigh's Ir (Perrault and Leigh 1989) and item placement ratio (Moore and Benbasat 1991). The value of these indices within each round of the Q-sorting are listed in Table 3.5.

Cohen's k as a measure of agreement can be interpreted as the proportion of joint judgment in which there is agreement after change agreement is excluded (Nahm et al. 2002), which is generally regarded as a conservative estimator of interrater reliability (Menor and Roth 2007). The score of Cohen's k greater than 0.65 indicates an adequate inter-judge agreement, occurring beyond chance (Moore and Benbasat 1991). Perreault and Leigh's (1989) interjudge agreement statistic (Ir) measures the observed proportion of agreement between judges (participants). The value is considered to be acceptable if greater than 0.65. Moore and Benbasat's (1991) item placement ratio assess content validity of the generated items and reliability of the proposed constructs. The value exceeds the recommended level of 0.70 is considered to be acceptable.

As shown in Table 3.5, Cohen's k, Ir and overall placement ratio exceed the cut-off in each four round, which shows a high degree of tentative reliability and construct validity. Overall, 14 items were deleted after the four rounds item-sorting analysis. Six items were deleted in first round, three items were deleted in second round, four items were deleted in third round, and one item was deleted in fourth round. See Tables 3.1, 3.2, 3.3 and 3.4 for a list of items deleted for each construct after Q-sorting (Table 3.5).

Taken together, the results of the literature review, interviews (n = 4), and Q-sorting (n = 8) provided tentative evidence of the reliability and validity of the NSD performance constructs and items. As a result of stage one, 22 measurement items capturing the four dimensions of NSD performance were retained for stage two of this study. NSD program refers to the portfolio of NSD projects the service organization has initiated within certain period (Menor and Roth 2007). The OECD recommends taking 3 year periods into account in innovation surveys, since innovation is a time dependent process (OECD-EUROSTAT 1997). Key informants were asked the following question: please indicate the extent to which you agree or disagree with the following statements as they pertain to all your new service development projects within the last 3 years. The items were evaluated on 7-point Likert scale responses from strongly disagree (1) to strongly agree (7).

Table 3.1 Measurement items of financial performance dimensions

Measurement items	Supporting literature
The new services achieved high overall profitability[c]	de Brentani (1989), Cooper et al. (1994), Storey and Easingwood (1996), Voss et al. (1992), Gray et al. (2002), Cooper and Kleinschmidt (1987), Griffin and Page (1993) and Story and Kelly (2001)
The payback period of investments in the new services was quite short[c]	Cooper and Kleinschmidt (1987) and Griffin and Page (1993)
The new services achieved important cost efficiencies for the company[b]	de Brentani (1989), Voss et al.(1992), Griffin and Page (1993) and Story and Kelly (2001)
The return on investment in the new services achieved or exceeded their objectives[c]	Cooper and Kleinschmidt (1987)
The new services exceeded their market share/sales/customer use level objectives[b]	de Brentani (1989), Cooper et al. (1994), Storey and Easingwood (1996), Voss et al. (1992), Cooper and Kleinschmidt (1987), Griffin and Page (1993) and Story and Kelly (2001)
The new services exceeded sales/client use growth objectives[c]	de Brentani (1989), Cooper et al. (1994), Storey and Easingwood (1996), Voss et al. (1992), Gray et al. (2002) and Story and Kelly (2001)
The new services achieved high relative sales/client use level[b]	de Brentani (1989), Cooper et al. (1994), Storey and Easingwood (1996), Gray et al. (2002) and Story and Kelly (2001)
The new services expanded the current market[a(1)]	de Brentani (1989)
The new services had a positive impact on the company's reputation[a(2)]	de Brentani (1989), Cooper et al. (1994), Storey and Easingwood (1996), Voss et al. (1992) and Gray et al. (2002)
The new services had a large market share[b]	de Brentani (1989), Cooper et al. (1994), Storey and Easingwood (1996), Voss et al. (1992) and Story and Kelly (2001)
The new services created new market opportunity[a(1)]	Cooper et al. (1994), Storey and Easingwood (1996), Voss et al. (1992), Cooper and Kleinschmidt (1987), Story and Kelly (2001)

[a(i)] These items were dropped during the i-th item-sorting analysis
[b] These items were dropped during the tentative confirmatory factor analysis
[c] These items were retained for the confirmatory factor analysis

Sample and procedures: Data for this study were gathered through enterprise questionnaire investigation in Chinese knowledge intensive business service across four sectors according to Miles's et al. (1995) industry classification: management consulting, engineering consultancy, advertising and software service. After questionnaire presetting for clarity and relevance through face-to-face interviews with executive in four firms, the survey procedures were organized as follows with the pre-tested questionnaires.

Firstly, we developed the initial samples of knowledge intensive business service firms through an exhaustive search on internet (The websites, such as www.0431-114.com, www.syyp.net, 0431.51ys.com and so on, include information about most

Table 3.2 Measurement items of customer performance dimensions

Measurement items	Supporting literature
The new services were clearly superior to competitors in terms of meeting client needs[a(3)]	Hipp and Grupp (2005)
The new services reduced the cost of clients[c]	de Brentani (1989)
The new services offered unique benefits to the clients not available elsewhere[c]	de Brentani (1989) and Voss et al. (1992)
The new services increased the productivity or product range of the clients[c]	Hipp and Grupp (2005)
The new services resulted in superior "service experience" to competitors[b]	de Brentani (1989), Voss et al. (1992) and Hipp and Grupp (2005)
The new services improved the satisfaction and loyalty of existing clients[c]	Cooper et al. (1994), Storey and Easingwood (1996), Griffin and Page (1993), Gray et al. (2002) and Story and Kelly (2001)
The new services accelerated speed to market for subsequent new services[a(1)]	Voss et al. (1992), Hipp and Grupp(2005) and Story and Kelly (2001)
The new services allowed a long-term client relationship to be built[a(2)]	Easingwood and Storey (1993b)

[a(i)] These items were dropped during the i-th item-sorting analysis
[b] These items were dropped during the tentative confirmatory factor analysis
[c] These items were retained for the confirmatory factor analysis

Table 3.3 Measurement items of internal process performance dimensions

Measurement items	Supporting literature
The systems (hardware, software, delivery systems) developed to launch these new services provided a basis for a better introduction of services in the future[c]	Cooper et al. (1994) and Storey and Easingwood (1996)
The new services increased motivation and productivity among employees[c]	Hipp and Grupp (2005)
The new services were more reliable, accurate and of consistent quality[c]	de Brentani (1989), Voss et al. (1992), Hipp and Grupp (2005) and Story and Kelly (2001)
The new services increased the general service delivery capability of our company[c]	Storey and Easingwood (1993a) and Hipp and Grupp (2005)
The new services extended or completed the service line[a(4)]	de Brentani (1989) and Storey and Easingwood (1993b)
The new services improved existing services[a(3)]	de Brentani (1989) and Storey and Easingwood (1993a)
The new services enhanced sales of other services[a(1)]	de Brentani (1989) and Voss et al. (1992)
The new services enhanced profitability of other services[a(1)]	de Brentani (1989), Cooper et al. (1994), Storey and Easingwood (1996) and Story and Kelly (2001)
The new services gave the company important competitive advantage[a(1)]	de Brentani (1989)

[a(i)] These items were dropped during the i-th item-sorting analysis
[b] These items were dropped during the tentative confirmatory factor analysis
[c] These items were retained for the confirmatory factor analysis

Table 3.4 Measurement items of learning and growth performance dimensions

Measurement items	Supporting literature
The expertise of developing and launching these new services led to an enhanced know-how for future innovation projects[c]	Blazevic and Lievens (2004)
The development of these new services created a general development expertise that eased the development and introduction of subsequent new services[a(3)]	Blazevic and Lievens (2004)
Employees learned a lot from the development and launch of these new services[c]	Blazevic and Lievens (2004) and Hipp and Grupp (2005)
The development of these new services augmented the company's knowledge of new markets or new technologies[c]	Maidique and Zirger (1985)
The development of these new services improved the new service development capability of our company[c]	Blazevic and Lievens (2004)
Through developing and launching of these new services, employees acquired knowledge from our clients[b]	Muller and Zenker (2001)
The development of these new services enhanced success rate of subsequent new services[a(2)]	Voss et al. (1992)
The development of these new services facilitated knowledge flows in our company[a(3)]	Blindenbach-Driessen and den Ende (2006)

[a(i)] These items were dropped during the i-th item-sorting analysis
[b] These items were dropped during the tentative confirmatory factor analysis
[c] These items were retained for the confirmatory factor analysis

Table 3.5 The degree of agreement between judges within each round of the Q-sorting

Theoretical construct	Cohen's k				Perreault and Leigh's Ir				Overall placement ratio (%)			
	1	2	3	4	1	2	3	4	1	2	3	4
Financial performance	0.90	0.87	0.91	0.94	0.94	0.93	0.95	0.97	91	94	100	100
Customer performance									88	92	92	100
Internal performance									72	100	83	90
Learning and growth performance									100	94	81	100
Sort round average overall placement ratio									87.5	95	89.29	97.83

Note 36 items in first round; 30 items in second round; 26 items in third round; 23 items in fourth round

firms in Changchun and Shenyang: classification, address and telephone). Secondly, their location was detailed and firms were classified. We listed all the office building where the sample firms located, and then classify these office buildings into eight groups according to their location in Changchun and Shenyang. Thirdly, we averagely divided the eight investigators into four groups and assign office building groups to each group of investigators. Finally, each group survey executive, who would like to finish the pre-tested questionnaires, of the knowledge intensive business service firms in the office buildings through face-to-face interviews in Changchun and Shenyang.

We have visited 600 firms and successfully surveyed 192 firms in which only firms with more than five employees were included in sample. The number of respondents from management consult was 44, engineering consultancy was 46, advertising was 49 and software service was 53 respectively. The response rate is about 30 %. Independent samples T test was carried out to identify significant differences between the firms in Changchun and Shenyang. No statistically significant differences were found in any of the questionnaire's items and the number of employees.

3.4.4 Analyses

Additional measurement item refinement: We employed CFA to tentatively estimate measurement models for each scale with errors allowed to freely correlate with each other, and identify several items for removal from the financial, customer, and learning and growth performance dimension scales to increase reliability and goodness of fit (see Tables 3.1, 3.2, 3.3 and 3.4), while ensuring continued content validity (Bollen 1989; Rosenzweig and Roth 2007). While Hensley (1999) advocates the use of as few items as possible to measure theoretical constructs, each of the construct dimensions was represented with four measurement items through this process, which is desirable from a structural equations modeling standpoint (Hinkin 1998). Next, CFA was then employed on the 16 retained items to assess measurement scale unidimensionality, reliability, convergent validity, discriminant validity and predictive validity for the four NSD performance dimensions.

Unidimensionality analysis: Measurement scales are considered to be unidimensional if the items in the scale measure a single construct (Menor and Roth 2007; Stratman and Roth 2002). Unidimensionality is a necessary condition for reliable and valid scales (Anderson and Gerbing 1991). We employed CFA on each of the scales separately to assess the unidimensionality of the NSD performance scales. Measurement models for each scale were estimated, with errors allowed to freely correlate with each other. In CFA, a goodness-of-fit (GFI) index of 0.90 or higher provides adequate evidence for unidimensionality, in addition to a nonsignificant Chi-squared statistic (Bollen 1989). We also employed adjusted goodness-of-fit index (AGFI), the comparative fit index (CFI), non-normed fit index (NNFI), and the

Table 3.6 Unidimensionality and reliability analyses of NSD performance scales

Construct scale	Items	χ^2(p-value)	GFI	AGFI	NNFI	CFI	RMSEM
Financial performance	4	1.96 (0.37)	0.99	0.97	1.00	1.00	0
Customer performance	4	1.17 (0.56)	0.99	0.98	1.01	1.00	0
Internal performance	4	1.03 (0.60)	1.00	0.99	1.01	1.00	0
Learning and growth performance	4	2.14 (0.34)	0.99	0.97	1.00	1.00	0.02

root mean square error of approximation (RMSEA) to examined absolute, incremental and parsimonious fit. The GFI, CFI and NFI Values higher than 0.90 are indicative of a good fit, while AGFI value higher than 0.80 suggests a good fit of the hypothesized model (Gefen et al. 2000). For RMSEA, a value less than 0.1 is considered a good fit, and a value less than 0.05 is considered a very good fit of the data to the research model (Steiger 1990). Table 3.6 lists the fit statistics/indices for the NSD performance scales.

As Table 3.6 shown, all of fit statistics/indices for the multi-item scales for the NSD performance dimensions were within the recommended level, indicating that they meet generally accepted criteria for unidimensionality.

Reliability and convergent validity analysis: Reliability refers to the attribute of consistency in measurement (Gregory 2005), which indicates to what extent the different items are coherent with each other and whether they can be used to measure a specific magnitude (Jerez-Gómez et al. 2005). To assess reliability, one way is to examine the internal consistency of the index using a coefficient known as Cronbach's alpha (Cronbach 1951), estimates greater than 0.70 are generally considered to meet the criteria for reliability. The other way is to employ composite reliability technique which employs confirmatory factor analyses (CFA) to derive a composite reliability index (P_C) (Fornell and Larcker 1981). The composite construct reliability assesses whether measurement items sufficiently represent their respective constructs (Bagozzi and Yi 1988). The items are sufficient in their representation of their respective construct if P_C exceeds the suggested 0.70 standard (Menor and Roth 2007).

Convergent validity relates to the degree to which multiple methods of measuring a variable provide the same results, which implies that if a measure is valid, it should yield the same results when utilized across different methods (O'Leary-Kelly and Vokurka 1998). One method of testing convergent validity is to use two completely different measurement approaches (Stratman and Roth 2002; Rosenzweig and Roth 2007). Because responses from both the two stages yielded similar results in terms of scale reliability and unidimensionality, some evidence for the convergent validity of our scales is established. Convergent validity can also be assessed by the magnitude and sign of the factor loadings of the measurement items (Menor and Roth 2007; Rosenzweig and Roth 2007; Stratman and Roth 2002).

As shown in Table 3.7, the resulting Cronbach's alpha and composite reliability indices for each of our NSD performance scales easily exceeded the 0.70 rule-of-thumb cut-off value, indicating that they meet generally accepted criteria for reliability. Each of our standardized path loadings is in the anticipated direction and

Table 3.7 NSD performance scales and items: standardized CFA path loadings and descriptive statistics

NSN performance scales and associated items[a]	Standardized path loading	Critical ratio[b]	Mean	S.D.
Financial performance	AVE = 0.54	Pc = 0.82	$\alpha = 0.83$	
The new services achieved high overall profitability	0.831	–	4.30	1.52
The payback period of investments in the new services was quite short	0.804	10.766	4.27	1.48
The return on investment in the new services achieved or exceeded their objectives	0.644	8.732	4.22	1.56
The new services exceeded sales/client use growth objectives	0.67	9.132	4.17	1.56
Customer performance	AVE = 0.49	Pc = 0.79	$\alpha = 0.79$	
The new services reduced the cost of clients	0.596	7.228	4.70	1.62
The new services offered unique benefits to the clients not available elsewhere	0.737	8.516	4.90	1.64
The new services increased the productivity or product range of the clients	0.772	–	4.85	1.65
The new services improved the satisfaction and loyalty of existing clients	0.669	7.999	5.22	1.58
Internal performance	AVE = 0.55	Pc = 0.83	$\alpha = 0.83$	
The systems (hardware, software, delivery systems) developed to launch these new services provided a basis for a better introduction of services in the future	0.673	8.736	5.20	1.59
The new services increased motivation and productivity among employees	0.788	–	4.83	1.63
The new services were more reliable, accurate and of consistent quality	0.758	9.737	4.76	1.48
The new services increased the general service delivery capability of our company	0.738	9.525	4.88	1.43
Learning and innovative performance	AVE = 0.62	Pc = 0.87	$\alpha = 0.87$	
The expertise of developing and launching these new services led to an enhanced know-how for future innovation projects.	0.777	9.832	5.50	1.53
Employees learned a lot from the development and launch of these new services	0.713	–	5.10	1.46
The development of these new services augmented the company's knowledge of new markets or new technologies	0.857	10.608	5.23	1.54
The development of these new services improved the new service development capability of our company	0.802	10.118	5.15	1.42

[a] These items represent the final refined measurement scales
[b] For a one-tailed test of significance, the critical ratio (CR) and associated p-values are as follows: CR = 1.28, p < 0.10; CR = 1.64, p < 0.05; CR = 2.33, p < 0.01; CR = 3.10, p < 0.001

is significantly different from zero at $p \leq 0.01$, i.e., the critical ratio is ≥ 2.33, which corroborates the existence of convergent validity. The average variance extracted values (AVE), which assess the amount of variance that is captured by the construct in relation to the amount of variance due to measurement error, and values equal or exceeding 0.5 indicate that the measures are reflective of the construct (Fornell and Larcker 1981). As reported in Table 3.7, the all AVE values exceed 0.5 except for that of customer performance. However the AVE value of customer performance is still marginally acceptable. The values of AVE indicate that a large amount of the variance is captured by each construct rather than due to measurement error. Together these results offer further statistical evidence that the items adequately reflect their corresponding constructs and that our NSD performance scales exhibit convergent validity.

Discriminant validity analysis: Discriminant validity refers to the uniqueness and the independence of the measure, i.e., the extent to which the measures are distinctly different from each other (Li et al. 2005). A Chi-square difference test can be employed to check for discriminant validity (Lattin et al. 2005; Li et al. 2005). The discriminant validity procedure entails running two CFAs on each pair of scales. In the first analysis of each pair of scales, the two constructs are allowed to freely correlate (i.e., unconstrained model). In the second, the correlation between the two constructs is set to 1 (i.e., constrained model). Using the resulting Chi-square statistics from each of these analyses, we conducted a Chi-square difference test for each pair of scales. The difference between the two Chi-square statistics is also Chi-square distributed, with degree of freedom equal to the difference in the number of degrees of freedom for the two models (Lattin et al. 2005). The results of Chi-square difference test are shown is Table 3.8.

As shown is Table 3.8, the p values associated with each of the Chi-square statistics are all less than 0.05, indicating strong support for the discriminant validity of each multi-item scales. Given the CFA results, the stage one Q-sorting routines appear to have been especially valuable in developing reasonably good measurement scales that held up to confirmation in our field study (Menor and Roth 2007; Rosenzweig and Roth 2007; Moore and Benbasat 1991).

The relationship between performance dimensions: As conceptual framework of NSD performance shown in Fig. 3.1, NSD performance is conceptualized as a multidimensional construct reflected by the following complementary first-order dimensions: financial performance, customer performance, internal performance, and learning and growth performance. To obtain a sounder judgment of the structural relationships between the four dimensions and the wider construct over which they lie, a second-order confirmatory factor analysis is made (Hair et al. 1999).

Following a competing models strategy, 16 models or structural equation systems were analyzed. In the first 15, a first-order confirmatory analysis was applied (Model: single dimension; Models 2–8: two dimensions; Models 9–14: three dimensions; Model 15: four dimensions). Model 16, on the other hand, is second-order model. Table 3.9 provides a summary of the results obtained from the

Table 3.8 Pairwise CFA tests of measurement scale discriminant validity

Construct scale pairs		Unconstrained model		Constrained model		χ^2 Difference[a]
		χ^2	d.f	χ^2	d.f	
Financial	Customer performance	24.150	19	165.307	20	141.157
performance	Internal performance	43.754	19	198.878	20	155.124
	Learning and growth performance	30.510	19	245.937	20	215.427
Customer	Internal performance	41.812	19	54.362	20	12.55
performance	Learning and growth performance	31.409	19	96.208	20	64.799
Internal performance	Learning and growth performance	23.116	19	140.456	20	117.34

[a] All χ^2 differences are statistically significant at the $p < 0.001$ level

confirmatory analysis to the covariance matrix of the items included in the NSD performance scale. The fit for each of the models has been evaluated using several indicators.

Table 3.9 shows that only Models 15 and 16 include acceptable values in all the indicators, with Model 16—the model proposed—being slightly better in terms of normed chi-aquare, AGFI, NNFI and the RMSEA values except for FGI. According to competing models strategy, the multidimensionality of the NSD performance construct is represented by a second-order factor, which is a parsimonious representation of the co-variation of the four performance factors (Edwards 2001). The second-order confirmatory factor analysis provides an empirical backing to the proposed conceptual model of NSD performance in Fig. 3.1.

Common method variance: As we use self-report organizational data, we have tested for problems associated with common method variance (Jerez-Gómez et al. 2005). Podsakoff and Organ (1986) review techniques that address method variance, including a procedure called the single factor approach. The logic of this approach is that if method variance accounts for the relations between two or more variables, a factor analysis should yield a single method factor. Exploratory factor analysis (EFA) revealed the first factor accounted for only 42.50 % of the variance; thereby a factor analysis did not yield a single method factor. McFarlin and Sweeny (1992) indicate that this approach has its weaknesses, although it provides useful information. They suggest combined it with a stronger confirmatory factor analysis which tests models that increase in complexity. As the results of confirmatory factorial analysis shown in Table 3.9, the least complex model examined, the single factor model, did not fit the data as well as more complex models. Thus, according to McFarlin and Sweeny's criterion (1992), the common method variance does not pose a problem in this study.

Predictive validity analysis: Predictive validity is a measure of how well antecedent constructs predict the hypothesized dependent variable (Stratman and Roth 2002; Rosenzweig and Roth 2007). Our conceptualization of measurement scale for new service development performance suggests that there may be a

Table 3.9 Summary results of confirmatory factor analysis: competing models

Models	χ^2	df	χ^2/df	GFI	AGFI	CFI	NNFI	RMSEM
M1: 1 dimension	509.059	104	4.895	0.721	0.635	0.723	0.680	0.143
M2: 2 dimensions (FP–CP) (IP–LGP)	458.816	103	4.717	0.720	0.631	0.738	0.695	0.139
M3: 2 dimensions (FP–IP) (CP–LGP)	348.367	103	4.265	0.746	0.664	0.770	0.733	0.131
M4: 2 dimensions (FP–LGP) (CP–IP)	410.46	103	3.985	0.764	0.689	0.790	0.755	0.125
M5: 2 dimensions (FP) (CP–IP-LGP)	323.297	103	3.139	0.806	0.744	0.849	0.824	0.106
M6: 2 dimensions (CP) (FP–IP-LGP)	493.564	103	4.792	0.725	0.637	0.733	0.689	0.141
M7: 2 dimensions (IP) (FP–CP-LGP)	468.911	103	4.553	0.736	0.651	0.750	0.708	0.136
M8: 2 dimensions (LGP) (FP–CP-IP)	370.207	103	3.594	0.790	0.723	0.817	0.787	0.117
M9: 3 dimensions (FP)(CP)(IP-LGP)	309.253	101	3.062	0.810	0.745	0.857	0.831	0.104
M10: 3 dimensions (FP)(IP)(CP–LGP)	271.569	101	2.689	0.835	0.778	0.883	0.861	0.094
M11: 3 dimensions (FP)(LGP)(CP–IP)	189.941	101	1.881	0.896	0.860	0.939	0.928	0.068
M12: 3 dimensions (CP)(IP)(FP–LGP)	396.347	101	3.924	0.770	0.690	0.798	0.760	0.124
M13: 3 dimensions (CP)(LGP)(FP–IP)	346.064	101	3.426	0.797	0.726	0.832	0.801	0.113
M14: 3 dimensions (IP)(LGP)(FP–CP)	352.033	101	3.485	0.792	0.719	0.828	0.796	0.114
M15: 4 dimension	174.313	98	1.779	0.903	0.865	0.948	0.936	0.064
M16: second-order confirmatory	176.391	100	1.764	0.902	0.866	0.948	0.937	0.063

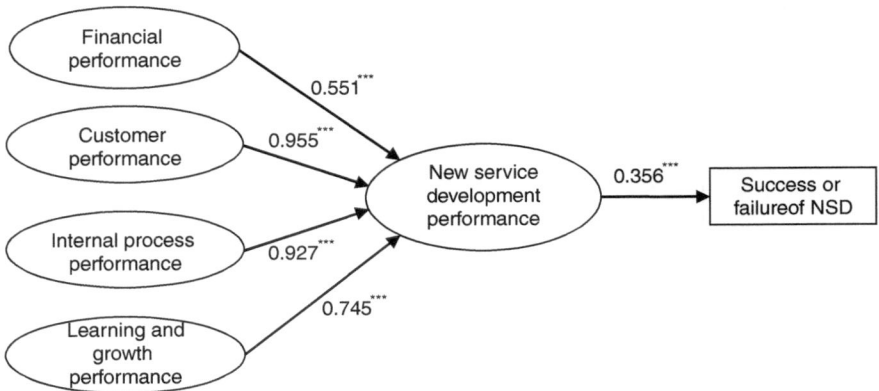

Fig. 3.2 Results of structural equation model (Fit measure:$\chi^2/df = 1.936$, GFI = 0.881, AGFI = 0.842, CFI = 0.929, NNFI = 0.916, RMSEM = 0.070 *** P < 0.001)

positive association between measurement scale and overall success/failure. In an attempt to establish predictive validity for the new service development performance scales, we asked respondents to provide us with four self-reported overall success/failure measures of their organization's NSD performance within the last 3 years. A success was globally defined as a portfolio of new services which had clearly met or exceeded the overall NSD development strategic objects for which they were developed, while failure had definitely not met the overall NSD development strategic objects for which they were developed. We employed one global measure of success/failure to rate NSD program with the last 3 years from 0 to ± 6, ranging from "Barely met our minimum strategic objects" [0] to "moderately successful(unsuccessful)"[±3]to "overriding success(failure)" [±6]. This approach has been used in other new service development management studies (de Brentani 1989).

On the base of the conceptual model of NSD performance (see Fig. 3.1), we implemented the structural equation modeling (SEM) using Amos 7.0. Figure 3.2 presents a graphical representation of the SEM models with their standardized regression coefficients for each construct. The standardized regression coefficients in Fig. 3.2 show that the four first-order dimensions has have a positive effect on NSD performance scales, which has a positive effect on overall success/failure. The results indicate that the proposed measurement scale for new service development performance could be use as an instrument to estimate the success or failure of NSD program. Consequently, the proposed measurement scale helps NSD managers in understanding and managing the performance of new service development program.

The results suggest NSD performances to a latent multidimensional construct inasmuch as its full significance lies beneath the four dimensions that go towards its makeup. Thus, NSD program should show a high degree of performance in each and every one of the four dimensions defined to be able to state that its NSD performance is high. These dimensions, called financial performance, customer performance, internal performance, and learning and growth performance, constitute our NSD performance structure model. Hence, managers that are to improve NSD performance must deal with four dimensions at the same time. Neglecting any one dimension seems to be a mistaken NSD strategy.

3.5 Conclusion

This chapter represents the development and validation of a new multi-item scale for NSD performance within BSC framework. The scales are validated using data from a survey of Chinese KIBS. The creation of this scale responds to the need expressed in the literature that there can be no effective commentary on which factors contribute most to NSD performance without a clearly understood conceptualization of success (Storey and Kelly 2001). Building on the balanced scorecard framework (Kaplan and Norton 1992, 1996), which is an effective mechanism for assessing success and failure of NSD program, NSD performance is conceptualized as a multidimensional construct reflected by the following complementary first-order dimensions: financial performance, customer performance, internal performance, and learning and growth performance. Each of these dimensions, in turn, is represented by a unidimensional multi-item scale. We utilized Menor and Roth's structured two-stage approach for our new scale development. This two-stage approach, utilizing two different data samples, allowed for rigorous statistical analyses to refine both individual measurement items and multi-item scales (Menor and Roth 2007). The scale has behaved well in the statistical analyses carried out to check for the presence of unidimensionality, reliability, convergent validity, discriminant validity and predictive validity. The second-order confirmatory factor analysis provides an empirical backing to the proposed conceptual model of NSD performance. NSD performance is considered to be a latent construct that lies under four dimensions which are also latent and measured using different observable variables.

The validated scale is useful both for academia and for practitioners. Academics can be used as a measurement instrument for further research on NSD performance. Though former studies into NSD performance have tended to different NSD performance dimension, they can be classified into subset of our proposed NSD performance dimensions. Thus, the proposed measures can facilitate work that allows the effect of different antecedents on NSD performance to be evaluated and make generalizable conclusions more easily drawn. The measurement scale proposed is useful for evaluating more complex models in which the effect of different antecedents on NSD performance can be analyzed. It will be a very fruitful area for future research. Practitioners could use the proposed scale as a diagnosing and benchmarking tool for NSD performance inside a particular firm or inside an industry. In addition, the validation process has shown other important implications for management: financial performance, customer performance, internal performance, and learning and growth performance are four dimensions that shape the concept of NSD performance. These four dimensions, based on the BSC framework (Kaplan and Norton 1992, 1996), can help the practitioners translating their NSD strategy into action. They are correlated and complementary, and have a strong positive relationship with NSD performance. Banker et al. (2000) also find that when non financial measures are included in compensation contract, managers more closely aligned their efforts to those measures, resulting in increased performance. Consequently, managers must therefore implement measures to improve the four dimensions in parallel.

The empirical analysis carried out, however, has revealed certain limitations to this study. The study has been conducted in only industry (KIBS) and in one national context (China). The study on a single industry within one national allows the context to be examined in greater detail and minimizes possible external influences on the NSD performance scale. However, it can also limit its external validity which refers to the generalization of research findings, either from a sample to a larger population or to settings and populations other than those studied (Lucas 2003). Due to the fact that measuring NSD performance with BSC framework have not been analyzed or considered previously, it would be worthwhile employing the NSD performance scale for testing in other national and industrial contexts, to establish its ultimate reliability and validity. In addition, common methods variance bias may be a potential problem in our study. Utilizing single informants in our survey data collection may create the potential for common methods variance bias in our data. Moreover, the propensity for respondents to try to maintain consistency in their responses to questions may also create the potential for common methods variance bias. Though the results from Harman's one-factor test (Podsakoff and Organ 1986) and confirmatory factor analysis which tests models that increase in complexity (McFarlin and Sweeny 1992) indicated that common-methods variance is not present to a significant degree in our data, but gives the potential for common methods variance bias in future research. Podsakoff et al.'s (2003) recommendations for controlling for common method variance in different research settings pave the way for interesting extension of this work. If the predictor and criterion variables can be measured from different sources, a logical extension of this research would be to test the model using survey data collected from multiple informants such as employees or customers. If the variables cannot be measured from different sources, then we recommend that researchers follow the Podsakoff et al.'s recommendations for how to select appropriate procedural and statistical remedies for different types of research settings (see Podsakoff et al. 2003).

The items of a test can be visualized as items that define what the research really wishes to measure (Gregory 2005). Our NSD competence construct as a multidimensional reflective indicator construct such that any changes to this set of performance indicators should not alter the conceptual domain of the NSD performance construct (Jarvis et al. 2003). As Griffin and Page (1993) argued that "If underlying dimensions could be identified, then regardless of which specific measures were used to quantify the firm's performance in a particular dimension, we might ultimately be able to help firms determine whether they were missing any aspects of measurement that would help provide them with a more balanced view of their performance", it will be a very fruitful area for future research to continuously refine of the NSD performance scale proposed and supported in this study through inclusion of new items or deletion of original items.

References

Ahn JH, Lee DJ, Lee SY (2006) Balancing business performance and knowledge performance of new product development: lessons from its industry. Long Range Plan 39(5):525–542

Alegre J, Lapiedra R, Chiva R (2006) A measurement scale for product innovation performance. Eur J Innov Manag 9(4):333–346

Amaratunga D, Baldry D, Sarshar M (2001) Process improvement through performance measurement: the balanced scorecard methodology. Work Study 50(5):179–188

Anderson JC, Gerbing DW (1991) Predicting the performance of measures in a confirmatory factor analysis with a pretest assessment of their substantive validities. J Appl Psychol 76(5):732–741

Andrews KZ (1996) Two kinds of performance measures. Harv Bus Rev 74(1):8–9

Auh S, Menguc B (2007) Performance implications of the direct and moderating effects of centralization and formalization on customer orientation. Ind Mark Manag 36(8):1022–1034

Avlonitis GJ, Papastathopoulou PG, Gounaris SP (2001) An empirically-based typology of product innovativeness for new financial services: success and failure scenarios. J Prod Innov Manag 18(5):324–342

Bagozzi RP, Yi Y (1988) On the evaluation of structural equation models. J Acad Mark Sci 16(1):74–94

Banker RD, Potter G, Srinivasan D (2000) An empirical investigation of an incentive plan that includes non-financial performance measures. Account Rev 75(1):65–92

Bhagwat R, Sharma MK (2005) Performance measurement of supply chain management: a balanced scorecard approach. Comput Ind Eng 53(1):43–62

Bititci US, Cavalieri S, Cieminski G (2005) Implementation of performance measurement systems: private and public sectors. Prod Plan Control 16(2):99–100

Blazevic V, Lievens A (2004) Learning during the new financial service innovation process antecedents and performance effects. J Bus Res 57(4):374–391

Blindenbach-Driessen F, den Ende JV (2006) Innovation in project-based firms: the context dependency of success factors. Res Policy 35(4):545–561

Bollen KA (1989) Structural equations with latent variables. Wiley, New York

Bose S, Thomas K (2007) Applying the balanced scorecard for better performance of intellectual capital. J Intellect Cap 8(4):653–665

Bryant L, Jones DA, Widener SK (2004) Managing value creation within the firm: an examination of multiple performance measures. J Manag Account Res 16(1):107–131

Churchill G (1979) A paradigm for developing better measures of marketing constructs. J Mark Res 16(1):64–73

Cohen J (1960) A coefficient of agreement for nominal scales. Educ Psychol Measur 20(1):37–46

Cooper RG, Kleinschmidt E (1987) New products: what separates winners from losers. J Prod Innov Manag 4(3):169–184

Cooper RG, Easingwood CJ, Edgett S, Kleinschmidt EJ, Storey C (1994) What distinguishes the top performing new products in financial services? J Prod Innov Manag 11(4):281–299

Cronbach LJ (1951) Coefficient alpha and the internal structure of tests. Psychometrika 16(3):297–334 de Brentani U (1989) Success and failure in new industrial services. J Prod Innov Manag 6(4):239–258

de Brentani U (2001) Innovative versus incremental new business services: different keys for achieving success. J Prod Innov Manag 18(3):169–187

Den Hertog P (2000) Knowledge-intensive business services as co-producers of innovation. Int J Innov Manag 4(4):491–528

Driva H, Pawar K, Menon U (2001) Performance evaluation of new product development from a company perspective. Integr Manuf Syst 12(5):368–378

Easingwood CJ, Storey CD (1993a) Marketplace success factors for new financial services. J Serv Mark 7(1):41–54

Easingwood CJ, Storey CD (1993b) Marketplace success factors for new financial services. J Serv Mark 7(1):41–54

Edwards JR (2001) Multidimensional constructs in organizational behavior research: an integrative analytical framework. Organ Res Methods 4(2):144–192

Fernandesa KJ, Rajab V, Whalleyc A (2006) Lessons from implementing the balanced scorecard in a small and size manufacturing organization. Technovation 26(5/6):623–634

Fitzgerald L, Johnston R, Brignall TJ, Silvestro R, Voss C (1991) Performance measurement in service businesses. CIMA, London

Fornell C, Larcker DF (1981) Evaluating structural equation models with unobservable variables and measurement error. J Mark Res 18(1):39–50

Gefen D, Straub D, Boudreau M (2000) Structural equation modeling and regression: guidelines for research practice. Commun AIS 4(7):1–76

Gray BJ, Matear S, Matheson PK (2002) Improving firm service performance. J Serv Mark 16(3):186–200

Gregory RJ (2005) Psychological testing: history, principles, and application. Peking University Press, Beijing

Griffin A, Page AL (1993) An interim report on measuring product development success and failure. J Prod Innov Manag 10(4):281–308

Hair JF, Anderson RE, Tatham RL, Black WC (1999) Multivariate data analysis (5th ed). Prentice-Hall, New York

Hart S (1993) Dimensions of success in new product development: an exploratory investigation. J Mark Manag 9(1):23–41

Hensley RL (1999) A review of operations management studies using scale development techniques. J Oper Manag 17(3):343–358

Hill M (1997) SPSS missing value analysis^{TM} 7.5. SPSS Inc, Chicago

Hinkin TR (1998) A brief tutorial on the development of measures for use in survey questionnaires. Organ Res Methods 1(1):104–121

Hipp C, Grupp H (2005) Innovation in the service sector: the demand for service-specific innovation measurement concepts and typologies. Res Policy 34(4):517–535

Hopwood AG (1974) Accounting and human behaviour. Accountancy Age Books, London

Jarvis CB, Mackenzie SB, Podsakoff PM (2003) A critical review of construct indicators and measurement model misspecification in marketing and consumer research. J Consum Res 30(2):199–218

Jerez-Gómez P, Céspedes-Lorente J, Valle-Cabrera R (2005) Organizational learning capability: a proposal of measurement. J Bus Res 58(6):715–725

Johne A, Storey C (1998) New service development: a review of the literature and annotated bibliography. Eur J Mark 32(3/4):184–251

Kaplan RS, Norton DP (1992) The balanced scorecard: measures that drive performance. Harv Bus Rev 70(1):71–79

Kaplan RS, Norton DP (1996) Translating strategy into action: the balanced scorecard. Harvard Business School Press, Boston

Lattin JM, Carroll JD, Green PE (2005) Analyzing multivariate data. China Machine Press, Beijing

Li S, Rao SS, Ragu-Nathan TS, Ragu-Nathan B (2005) Development and validation of a measurement instrument for studying supply chain management practices. J Oper Manag 23(6):618–641

Lillis AM (2002) Managing multiple dimensions of manufacturing performance: an exploratory study. Account Organ Soc 27(6):497–529

Lingle JH, Schiemann WA (1996) From balanced scorecard to strategic gauges: is measurement worth it? Manag Rev 85(3):56–62

Lipe MG, Salterio SE (2000) The balanced scorecard: judgmental effects of common and unique performance measures. Account Rev 75(3):283–298

Lucas JW (2003) Theory-testing, generalization, and the problem of external validity. Socioll Theory 21(3):236–253

Maidique MA, Zirger BJ (1985) The new product learning cycle. Res Policy 14(6):299–313

Mattoo A (2003) China's accession to the WTO: the services dimension. J Int Econ Law 6(2):299–339

McCracken MJ, McIlwain TF, Fottler MD (2001) Measuring organizational performance in the hospital industry: an exploratory comparison of objective and subjective methods. Health Serv Manag Res 14(4):211–219

McFarlin DB, Sweeney PD (1992) Distributive and procedural justice as predictors of satisfaction with personal and organizational outcomes. Acad Manag J 35(3):626–637

Menor LJ, Roth AV (2007) New service development competence in retail banking: construct development and measurement validation. J Oper Manag 25(4):825–846

Miles I, Kastrinos N, Flanagan K, Bilderbeek R, Hertog P, Huntink W, Bouman M (1995) Knowledge intensive business services: users, carriers and sources of innovation, Rappon pour DG13 SPRINT-EIMS, March

Moore GC, Benbasat I (1991) Development of an instrument to measure the perceptions of adopting an information technology innovation. Inf Syst Res 2(3):192–222

Muller E, Zenker A (2001) Business services as actors of knowledge transformation: the role of KIBS in regional and national innovation systems. Res Policy 30(9):1501–1516

Nahm AY, Solis-Galvan LE, Rao SS, Ragu-Nathan TS (2002) The Q-sort method: assessing reliability and construct validity of questionnaire items at a pre-testing stage. J Mod Stat Methods 1(1):114–125

Nakos G, Brouthers LE, Brouthers KD (1998) The impact of firm and managerial characteristics on small and medium-sized greek firms' export performance. J Glob Mark 11(4):23–47

Neely A (1998) Measuring business performance. Economist Books, London

O'Leary-Kelly SW, Vokurka RJ (1998) The empirical assessment of construct validity. J Oper Manag 16(4):387–405

OECD (1997) National innovation systems. OECD Publications, Paris

OECD-EUROSTAT (1997) The measurement of scientific and technological activities. Proposed guidelines for collecting and interpreting technological data, Oslo Manual. OECD, Paris

Ottenbacher M, Gnoth J, Jones P (2006) Identifying determinants of success in development of new high-contact services: Insights from the hospitality industry. Int J Serv Ind Manag 17(4):344–363

Pangarkar AM, Kirkwood T (2008) Strategic alignment: linking your learning strategy to the balanced scorecard. Ind Commer Train 40(2):95–10

Perrault WD, Leigh LE (1989) Reliability of nominal data based on qualitative judgements. J Mark Res 26(2):135–148

Podsakoff PM, Organ DW (1986) Self-reports in organizational research: problems and prospects. J Manag 12(4):531–544

Podsakoff PM, MacKenzie SB, Lee JY, Podsakoff NP (2003) Common method biases in behavioral research: a critical review of the literature and recommended remedies. J Appl Psychol 88(5):879–903

Rosenzweig ED, Roth AV (2007) B2B seller competence: construct development and measurement using a supply chain strategy lens. J Oper Manag 25(6):1311–1331

Sampson P (1970) Can consumers create new products?. Mark Res Soc J 12(1):40–52

Scholey C (2005) Strategy maps: a step-by-step guide to measuring, managing and communicating the plan. J Bus Strat 26(3):12–19

Steiger JH (1990) Structural model evaluation and modification: an internal estimation approach. Multivar Behav Res 25(2):173–180

Storey CD, Easingwood CJ (1996) Determinants of new product performance: a study in the financial services sector. Int J Serv Ind Manag 7(1):32–55

Storey C, Kelly D (2001) Measuring the performance of new service development activities. Serv Ind J 21(2):71–90

Strambach S (1997) Knowledge-intensive services and innovation in Germany, report for TSER project, University of Stuttgart

Strambach S (2001) Innovation process and the role of knowledge-intensive business services. In: Kulicke KM, Zenker A (eds) Innovation networks—concepts and challenges in the European perspective. Physica-Verlag, Heidelberg, pp 53–68

Stratman JK, Roth AV (2002) Enterprise resource planning (ERP) competence constructs: two-stage multi-item scale development and validation. Decis Sci 33(4):601–628

Van der Stede WA, Chow CW, Lin TW (2006) Strategy, choice of performance measures, and performance. Behav Res Account 18(1):185–205

Voss C, Johnston R, Silvestro R, Fitzgerald L, Brignall T (1992) Measurement of innovation and design performance in services. Design Manag Jl 3(1):40–46

Wall TD, Michie J, Patterson M (2004) On the validity of subjective measures of company performance. Pers Psychol 57(1):95–118

Chapter 4
Identifying Determinants of Top Performing New Service Development Activities

4.1 Introduction

It is intense competition, deregulation of markets, emergence of globalization, greater customer sophistication and changes in technology that have made the successful development of new services a key to success for many firms (de Brentani 1989; Stevens and Dimitriadis 2005; Storey and Kelly 2001a, b; Peaks and Riihela 2004). Service firms are becoming increasingly innovative and are playing an ever increasing role in the knowledge-driven economies of both the developed and the developing countries (Howells 2003). New service development (NSD) has emerged as an important research topic in service marketing, innovation management and operation management (Johne and Story 1998; de Brentani 1989; Menor and Roth 2007).

Since the seminal work of de Brentani (1989), there has been considerable empirical research into the determinants of NSD performance in the developed countries (Cooper et al. 1994; Johne and Story 1998; Menor and Roth 2007). While each chapter is interesting, there are some problems in the current researches. Firstly, it has become apparent from these research findings that micro-level variables and meso-level variables, such as particular characteristics of firm and the type of industry, indeed influence the NSD performance. Though some macro-level variables, such as trajectories and institution, are also driving force or barriers behind service innovation (Sundbo and Gallouj 1998; Howells 2003), few empirical researches have investigated these factors' influence on NSD performance. An empirical research of these macro-level variables would provide important insights in the dynamics of innovation processes and open up opportunities for improving innovation activities and competitiveness of service industries. Consequently, we need to look more closely at macro-level variables driving service innovation.

S. Liu, *Innovation Management in Knowledge Intensive Business Services in China*, 43
SpringerBriefs in Business, DOI: 10.1007/978-3-642-34676-7_4,
© The Author(s) 2013

Secondly, much of current literature is limited to production-intensive services and supplier-dominated service such as banks, insurances, wholesale, telecommunications and hospitality industry (Edgett and Parkinson 1994; de Jong and Vermeulen 2003; Ottenbacher et al. 2006). Impediments to service innovation differ between service industries (Preissl 1998), which implies that these findings cannot be generalized to other service sectors. Lovelock and Gummesson (2004) suggest abandoning services as a general category altogether and focus research on specific service subfields. Empirical studies are both necessary and appropriate for researchers to explore a specific service industry (Cowell 1988; Easingwood 1986). The rise of knowledge economy has put knowledge intensive business services (KIBS) high on the research agenda (Kuusisto and Viljamaa 2004). Since 2000, Chinese KIBS had been developed rapidly (Liu 2008). However, this sector has been increasingly challenged by international competition for the Chinese accession to the WTO. KIBS is increasingly regarded as having a substantial role in firm innovation and as fairly homogenous when studying innovation (de Jone and Kemp 2003), moreover, a large share of innovative efforts in KIBS is related to the development of new services (den Hertog 2000; Muller and Zenker 2001). This study investigates whether the findings from production-intensive service and supplier-dominated service sectors are also applicable in knowledge intensive business service; and thereby will expand the scope of NSD research.

Thirdly, the previous studies on the determinants of NSD performance focused on project level (de Brentani 1989; Cooper et al. 1994; Ottenbacher et al. 2006). de Brentain (2001) argue that firms undertaking new service development are likely to be involved in, not one, but a portfolio of different types of projects. NSD program refers to the portfolio of NSD projects the service organization has initiated within certain period (Menor and Roth 2007). Undue focus on project performance in the product development literature has resulted in neglect of wider strategic considerations (Johne and Storey 1998). In this chapter, we measure NSD performance on program-level (Storey and Kelly 2001a, b) and investigate the determinants of NSD performance at program level.

Finally, the literature on NSD has been largely derived from research in the western world. Country-specific regulatory issues create country-specific impediments to service innovation (Preissl 1998). However, there is little literature on that in China. Managing NSD in China requires a set of complexities less common in the developed economies of the Western countries. In China, research and practice in NSD is at the exploratory stage. It requires new knowledge and is likely to need departures from existing tracks. This chapter aims to contribute to a better appreciation and understanding of the determinants of NSD performance in China, and will highlight research and practices surrounding NSD management in China.

The objective of this chapter is to gain a better understanding of how top performing NSD activities within KIBS are implemented in Chinese context, and furthermore investigate if the influence of macro-level factors provides additional explanation of the NSD performance.

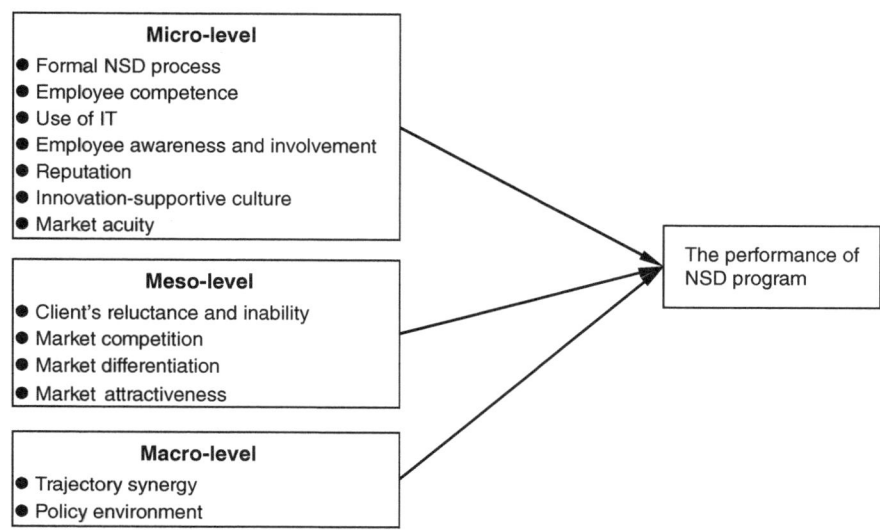

Fig. 4.1 Research conceptual model

4.2 Research Conceptualization

Based on our research objective, the dependent variable to be explained is the NSD program performance. To explain the performance of NSD program, we categorized the critical dimensions that influence the performance of NSD program into three categories, which relate to: micro-level dimension, mesco-level dimension and macro-level dimension. Our conceptual model of determinants of the performance of NSD program is depicted in Fig. 4.1.

While micro-level and mesco-level dimensions have long been consider by researcher as determinants of success and failure of new service development (Johne and Story 1998), with few exceptions (Sundbo and Gallouj 1998; Howells 2003), macro-level dimensions have been the topic of very little research on NSD, which are still limited to western countries. We provide below introduction of dimensions in each level and review the relevant scholarly literature.

4.2.1 Macro Level Dimensions

Literature suggest two macro factors that affect the NSD performance: policy environments and trajectories synergy (Lundvall and Borrás 2005; Preissl 1998; Sundbo and Gallouj 1998; Dosi 1982). The policies that encourage firm innovation are defined as the combined actions directed towards increasing the amount and intensity of innovation activities (Lundvall and Borrás 2005). Preissl (1998) believe that regulation as well as legislation and administrative rules play an important role and may encourage or discourage innovative activities. Howells (2003) identifies that the four institution barriers that obstacles to contemporary knowledge intensive business service innovation are (1) knowledge appropriability and intellectual property rights; (2) the difficulty in valuing and financing knowledge assets; (3) trade and internationalization; and (4) government support for innovation. The researches have shown that these innovation policies are the key aspects that affect innovation performance.

Trajectories are ideas and logics that are diffused through the social system which includes many and difficult identifiable actors (Sundbo and Gallouj 1998). In a report from the project Services in Innovation, Innovation in Services–Services in European Innovation Systems (SI4S), Sundbo and Gallouj (1998) identify five types of trajectories (1) Service professional trajectories: mean methods, general knowledge and behavioral rules that exist within the different service professions. (2) Managerial trajectories: refers to general management ideas or ideas for new organizational forms such as motivational systems, BPR, service management etc. (3) Technology trajectories: refers to new logics for using technology that generally influences service products and production processes. (4) Institutional trajectory: describes the general trend of evolution of regulations and political institutions. (5) Social trajectory: displays the evolutions of general social rules and conventions. The importance of trajectories is however not the actors, but the ideas and the logic behind the ideas (Sundbo and Gallouj 1998). The new services fitting with these trajectories may have better performance than those unfitting.

4.2.2 Meso Level Dimensions

Most studies on service innovation include two types of meso factors: client characteristic and market environments. According to Sundbo and Gallouj (1998) and Storey and Easingwood (1996), client characteristics, which impede service innovations, can be classified into three groups: socio-psychological factors, economic factors and understanding. Socio-psychological factors include conservation and reluctance to engage in innovative activities and accept new services (Preissl 1998). It is economic factors that potential clients do not see the value of the service for their company or think they can provide the same functions cheaper by themselves (Preissl 1998). Customers are more likely to purchase a new service if they understand the service, thus new service development success is also more likely if the customer understands the service offering (Storey and Easingwood 1996). Storey and Easingwood (1996) argue that customer understanding of a new service is helped by previous experience with the product class and by the relative simplicity of the products developed. This is more likely for existing customers of the company who are already familiar with the type of services that the company offers.

Market environments, which include market competition, market differentiation and market attractiveness, describe the characteristics of the market for the new services. Market attractiveness, which refers to high-growth markets, high-dollar-volume markets or margin market, is one of the factors discriminating between successful and less successful new service project (Ottenbacher et al. 2006). Cooper and de Brentani (1991) argue that services targeting large and high-growth markets are more successful than those aiming at smaller, slower-growth areas. Market attractiveness has often been found to be important for new service success (Ottenbacher et al. 2006, Johne and Storey 1998). However, the degree of competition, such as aggressive competition, price competition, similar competitive offerings, frequent product introductions/modifications and dominant competitors with large market share (s), appears not to be related to performance (Johne and Storey 1998; Cooper and de Brentani 1991) because competition is a fact of life in most service industries (Johne and Storey 1998). Firms operating in markets with no price competition are usually following a differentiation strategy (Porter 1980). Differentiation makes it more difficult for competitors to copy a service offering (MacMillan et al. 1985; Miles et al. 1995). Intellectual property protect is one of problems that obstacles to contemporary services innovation (Howells 2003), thus differentiation market may give a natural rise to the success of service innovation.

4.2.3 Micro Level Variable

Service innovations do not happen by chance, but formal NSD processes. de Brentani (1989) suggest that the effective development of new services require formal processes and practices like those typically found in new product development. The formal development process include formal idea screening, "drawing-board" design approach, thorough and realistic business analysis, formal monitoring/evaluation, formal and extensive launch and formal development team (de Brentani 1989, 1991; Edgett 1994; Cooper and de Brentani 1991; Edgett and Parkinson 1994; Storey and Easingwood 1998; Atuahene-Gima 1996a, b; Ottenbacher et al. 2006). Since firms that conduct more rigorous forms of service development activity are found to be more successful at developing and launching new services (Edgett and Parkinson 1994), firms need a formal new service development process (de Brentani 1989).

Though technology is not always a service innovation dimension, in practice, there is a wide range of relationships between technology and service innovation (den Hertog 2000). Lee et al. (2003) argue that "the key factor of convergence of innovation patterns appears to be IT technologies, which are vastly applied to the development of various other services. Integration of IT into many knowledge intensive services has led to a new paradigm of service innovation". IT use for NSD purposes reflects greater NSD competence (Menor and Roth 2007) and is major sources of innovation (den Hertog 2000). IT, though not the only relevant technology, is particularly pervasive in service innovation (den Hertog 2000).

IT can be used to speed up the introduction of new services, identify and diagnose customer needs and share information and facilitate communication and information flow (Menor and Roth 2007).

Edgett and Parkinson (1994) argue that "organizationally, successful new services exhibited strong inter-functional co-operation and co-ordination with development personnel fully aware of why they were involved and the importance of the new service's benefits to the company". Employee awareness is pivotal to the success of new service development. If employees don't understand the impact of their own actions, a company's efforts to develop new service will be futile. Strong internal marketing, to raise awareness within the company, is also needed in the new service development (Edgett 1994). Intra-organizational involvement has shown important influence on success of the new services development (Edgett and Parkinson 1994). Employee awareness is one of the foundations to new service success (Cooper et al. 1994).

Services are often thought to depend on people, and particularly their skills and knowledge, to an unusually high degree (Tether and Hipp 2002). Employees are the ones who have come up with ideas for new or improved services, and turn these into successful innovation (de Jong and Vermeulen 2003); thereby they are at the heart of the innovation process (Van de Ven 1986). Gallouj and Weinstein (1997) argue: 'One of the major features of service activities is undoubtedly the fact that the "technologies" involved usually take the form of knowledge and skills embodied in individuals (or teams) and are implemented directly when each transaction occurs, rather than in physical plant or equipment'. To archive the top NSD performance, new service should be produced by trained/skilled employees (Cooper and de Brentani 1991; Storey and Easingwood 1996, 1998). They, beside ability to interpret the market and an excellent view on unsatisfied client needs, must be creative in dealing with unique situation, well-educationed, well qualified for their tasks (de Brentani 1989, 1991; Ottenbacher et al. 2006; de Jong and Vermeulen 2003; Tsai et al. 2005).

Company's reputation measures the degree of confidence the customer as in the company and its products, how reliable the firm is, and its perceived expertise in its field (Storey and Easingwood 1998). Services are primarily intangible, which means that buyers frequently rely on company reputation when evaluating a new service, achieving a positive corporate image for superior quality, for trustworthiness or expertness can be an important form of success for service firms (de Brentani 1989). Because consumers may be willing to pay a price premium for a service they can trust, company's reputation are highly significant for profitability and sales (Storey and Easingwood 1998).

Innovation-supportive culture is defined as a firm's social and cognitive environment, the shared view of reality, and the collective belief and value systems reflected in a consistent pattern of behaviors among participants, which fostering expectations and guidelines for member creativity, experimentation and risk taking (Jassawalla and Sashittal 2002). Developing innovative services requires an environment that encourages and supports openness, creativeness and "stepping out" beyond the norm (Anderson and West 1998; de Brentani 2001). Cooper and Kleinschmidt (1995) note that innovative supportive culture include support team work, the emergence of product champions, support in terms of rewards, risk taking, autonomy in the treatment of failures, encouragement of employees to development their own ideas, and making venture capital or seen money available for internal projects. Chandler et al's research (2000) indicate a positive association of an innovation-supportive culture on firm earnings under conditions of rapid environmental change. Most researchers agree that supportive culture can enhance innovative performance and behavior (Menor and Roth 2007; Den Jong and Kemp 2003).

Menor and Roth (2007) define market acuity as the ability of the service organization to see the competitive environment clearly and to anticipate and respond to customers' evolving needs and wants. Market acuity is valuable to NSD success because it requires that the organization continuously collect information on customer needs and competitor capabilities, and uses this information to create new services that deliver superior customer value (Lucas and Ferrell 2000). Customer- and market- focus is one of the dominant success ingredients for top performing new service products (Cooper et al. 1994). Sunbo (1997) also argue that consideration of customers, competitors and market possibilities is the foundation for successful innovation efforts. Market acuity has been linked to business performance (Menor et al. 2001).

4.3 Research Methods

4.3.1 Pretests and Measures

The previous NSD literature provides a large pool of existing valid items to use (see e.g. de Brentani 1989; Cooper et al. 1994; Johne and Story 1998; Menor and Roth 2007). The items were prepared in English and then translated to Chinese by a panel including professors specialize in service and innovation management and Chinese professional translators. After pre-testing with four executes in KIBS, the final questionnaire contained 94 items, apart from those gathering demographic and performance information. Two English translators were employed to independently translate the 94 items back into English. Only minor inconsistencies occurred and resolved through communication with two translators. Our translation process was similar to the Avlonitis et al's (2001) parallel translation/double translation method; hence the translation is adequate so that these items can also be useful as that in the Western counties. The items were evaluated on 7-point Likert scale responses from strongly disagree (1) to strongly agree (7).

4.3.2 Sample and Procedure

Data for this study were gathered through enterprise questionnaire investigation in Chinese knowledge intensive business service across four sectors according to Miles et al. (1995) industry classification: management consulting, engineering consultancy, advertising and software service. After questionnaire presetting for clarity and relevance through face-to-face interviews with executive in four firms, the survey procedures were organized as follows with the pre-tested questionnaires.

Firstly, we developed the initial samples of knowledge intensive business service firms through an exhaustive search on internet (The websites, such as www.0431-114.com, www.syyp.net, 0431.51ys.com and so on, include information about most firms in Changchun and Shenyang: classification, address and telephone). Secondly, their location was detailed and firms were classified. We listed all the office building where the sample firms located, and then classified these office buildings into eight groups according to their location in Changchun and Shenyang. Thirdly, we averagely divided the eight investigators into four groups and assign office building groups to each group of investigators. Finally, each group survey executive, who would like to finish the pre-tested questionnaires, of the knowledge intensive business service firms in the office buildings through face-to-face interviews in Changchun and Shenyang.

We have visited 600 firms and successfully surveyed 192 firms in which only firms with more than five employees were included in sample. The number of respondents from management consult was 44, engineering consultancy was 46, advertising was 49 and software service was 53 respectively. The response rate is about 30 %.

4.3.3 Performance Measure

Kaplan and Norton (1992) develop the balanced scorecard (BSC) to complement traditional financial measures of business unit performance. Storey and Kelly (2001a, b) suggest that the BSC was suitable framework for measuring NSD performance at program level. According to the framework of balanced scorecard, NSD performance contains a diverse set of performance measures, spanning financial performance, customer relations, internal business processes, and the organization's learning and growth activities (Kaplan and Norton 1992, 1996). In an attempt to categorize top performing NSD activities and bottom performing ones on the base of the aggregate mean scores of measures that examined NSD performance, we asked respondents to provide us with self-reported measures of their NSD performance (see Table 3.7 for details).

The OECD recommends taking 3 year periods into account in innovation surveys, since innovation is a time dependent process (OECD-EUROSTAT 1997). Consequently, key informants were asked to indicate the extent to which agree or disagree (from strongly disagree (1) to strongly agree (7)) with the performance measure items as they pertain to all new service development projects within the last 3 years.

Like those of western managers (Ottenbacher et al. 2006), the pretest with Chinese managers also showed that they were reluctant to discuss unsuccessful in NSD activities. According to the approach widely used in other NSD studies (e.g. Cooper et al. 1994; Ottenbacher et al. 2006), we looked at the top half NSD performing NSD activities and contrasted them to the bottom half ones. We use median 4.938 as cut-point. Top and bottom performing NSD activities were defined, with NSD with a score above 4.938 considered top performing and NSD scoring below 4.938 considered bottom performing.

4.3.4 Identification of the Predictive Factors for Top Performing NSD Activities

An exploratory factor analysis using principal components analysis with varimax rotation was carried out to determine how the raw items were linked to their underlying factors. Items were excluded from the preliminary version if they had low factor loadings (<0.5) or loaded lower on its associated factor than on any other construct; thereby we can confirm the convergent validity and discriminant validity of the factors (Chau and Tam 1997). The results of Kaiser-Mayer-Olkin's Measure (KMO) of sampling adequacy test (0.879) and Bartlett tests (significance level <0.001) shows that the remained data meet the fundamental requirements for factor analysis (Bartlett 1954; Kaiser 1974). The factor analysis highlighted the existence of 13 factors with eigenvalues greater than 1.0. These factors accounted for 72.55 % of variance. None of the items cross-loaded on multiple factors supported unidimensionality (Bagozzi 1980). According to Nunnally (1978), acceptable Cronbach's a coefficients start at 0.70. In our study, all Cronbach's coefficients except one were greater than 0.70, which indicates reasonably high reliability. The exceptions are market differentiation with Cronbach's a = 0.54, which still meets the minimum standard of 0.5 as suggested by Nunnally (1978). Table 4.1 provides the scale items, the exploratory factor analysis factor loadings of the items utilized in this study and Cronbach's coefficient of these factors.

4.4 Finds

To predict the ability of the variables (represented by factor scores) to discriminate between top performing NSD activities and bottom performing ones, we employed a two-group Wilk's lambda stepwise discriminant analysis which is particularly suited to this situation, as the dependent variable is non-metric while the independent

Table 4.1 Descriptive factors of new service characteristic

Variables	Factor loadings
Trajectory synergy Cronbach's alpha = 0.911	
New services fitted with the general trend of evolution of regulations and political institutions	0.678
New services fitted with the evolutions of general social rules and conventions	0.712
New services fitted with general management ideas or ideas for new organizational forms such as motivational systems, BPR, service management etc.	0.751
New services fitted with methods, general knowledge and behavioral rules that exist within the different service professions	0.715
New services fitted with new logics for using technology that generally influences service products and production processes	0.562
Policy environment Cronbach's alpha = 0.843	
Company rarely received government support as to new service development needs	0.759
Service innovation lacked adequate protection to intellectual property rights	0.783
Government had no policy to encourage our firm to investigate new forms of service delivery across much wider markets	0.829
It was very difficult for our new service development upon an absence of venture capital and the unsympathetic attitude of financial organizations and banks towards service-based firms	0.756
Market competition Cronbach's alpha = 0.765	
New services targeted extremely aggressive competition	0.499
New services targeted intensive price competition	0.840
There were very similar competitive offerings	0.771
Market differentiation Cronbach's alpha = 0.544	
Our market was characterized by services that other firms can hardly imitate	0.698
Clients were insensitive to small price increases	0.663
Market attractiveness Cronbach's alpha = 0.795	
New services targeted large dollar volume market	0.614
New services targeted high margin market	0.512
New service targeted high growth market	0.805
Innovation-supportive culture Cronbach's alpha = 0.911	
Our firm encouraged entrepreneurial efforts and was accepting of risk-taking efforts	0.828
The managers frequently involved their staff in important decision-making processes.	0.771
Employee carried out their task without excessive supervision	0.781
Enough resources (time, money and people) were used for the new service development activities	0.688
Senior management placed strong and visible support behind NSD	0.599
Employee provided practical support for new ideas and their application	0.556

(continued)

Table 4.1 (continued)

Variables	Factor loadings
Employee awareness Cronbach's alpha = 0.886	
Employees involved in NSD were aware of the potential benefit new services would be to the company	0.554
Employees were informed prior to product launch	0.681
Employees involved in NSD knew why they were involved	0.644
The marketing cases were well made and understood at all levels in the company	0.713
There was a high level of awareness within the company that these new services were being developed	0.683
Employees from other functional groups were included in the development process as early as possible	0.726
Formal NSD process Cronbach's alpha = 0.913	
The new service idea had to pass an initial screening process before funds were allocated to it	0.653
A detailed written description of the service concept was developed very soon after the new service idea was accepted	0.778
In the pre-development stage, commercial and technical feasibility were investigated	0.717
During the various stages in the development process, formal monitoring/ evaluation program employed	0.767
The actual development process became more formal within the company as the development process evolved	0.778
Services were developed by a formal development team	0.660
Employee competence Cronbach's alpha = 0.86	
Employees were creative in dealing with unique situation	0.629
Employees were well-educationed	0.672
Employees were well qualified for their tasks	0.754
Employees had ability to interpret the market	0.702
Employees had an excellent view on unsatisfied client needs and were the first to recognize opportunities for innovation	0.644
Employees were extensively trained	0.712
Use of IT Cronbach's alpha = 0.899	
IT was used to speed up the introduction of new services	0.712
IT was used to identify and diagnose customer needs	0.682
IT was used to share information that coordinates new service/product development activities	0.743
Communication flow within team was facilitated through IT-based channels	0.734
Reputation Cronbach's alpha = 0.927	
Company had a strong commitment to our customers	0.710
Company had reputation for expertise	0.765
Company's reputation were important for the trial of new service	0.695
Company had reputation for quality	0.750
Customer satisfied with previous service	0.760

(continued)

Table 4.1 (continued)

Variables	Factor loadings
Customers were very loyal to existing service relationships	0.660
Client's reluctance and inability Cronbach's alpha = 0.903	
Clients were reluctant to accept new services	0.783
Clients were reluctant to engage in innovative activities	0.832
Clients feared the changes involved in their organizations	0.762
Clients could provide the same functions cheaper by themselves	0.702
Clients lacked competence to use new services	0.786
Service concept were difficult for clients to understand	0.773
Market acuity Cronbach's alpha = 0.853	
Company actively sought out information about our company's business environment	0.567
New offerings were designed based on information actively collected on evolving market shifts and customer demands for these offerings	0.574
Company used collected information to respond quickly to changes in the competitive environment	0.549
Customers, both internal and external, were viewed as potential and valuable sources of new offering ideas and opportunities	0.566
Company often collected information on our competitors' service offerings	0.695

Table 4.2 Results of the multiple discriminant analysis

Factor	Discriminant function coefficients	
	Unstandardised	Standardised
Formal NSD process	0.392 (4)	0.385 (4)
Employee competence	0.858 (1)	0.782 (1)
Trajectory synergy	0.292 (5)	0.290 (5)
Reputation	0.437 (3)	0.428 (3)
Innovation-supportive culture	0.598 (2)	0.574 (2)
Market attractiveness	0.284 (6)	0.282 (6)

Note numbers in parentheses represent the ranking of the coefficient from high (1) to low (6)

variable is metric (Edgett and Parkinson 1994). The two-group Wilk's lambda stepwise discriminant analysis can eliminate variables that are not good discriminators and considers the differences between groups as well as the cohesiveness or homogeneity within groups (Klecka 1980), and has been widely used in identifying determinants of success in development of new services (Edgett and Parkinson 1994; Ottenbacher et al. 2006).

Table 4.2 shows the six factors that best discriminated between top performing NSD activities and bottom ones. According to the ranking of the coefficient from high to low, these factors include employee competence, innovation-supportive culture, reputation, formal NSD process, trajectory synergy and market attractiveness. Seven factors are excluded from the final stepwise solution, indicating

Table 4.3 Discriminant function and related statistics

Group	Function	Related statistics	
Top performing	0.747	Box's M	77.750
Bottom performing	−0.812	Eigenvalue	0.614
		Canonical Correlation	0.617
		Wilks' Lambda	0.620
		Chi-square	89.468
		Degrees of freedom	6
		Significance	0.000

Table 4.4 Classification matrix

Group	Actual number of cases	Classification results Predicted group membership	
Stepwise			
Top performing	100	85 (85.00 %)	15 (15 %)
Bottom performing	92	25 (27.17 %)	67 (72.83 %)
Correctly classified		79.17 %	
U-method			
Top performing	100	82 (82.00 %)	18 (18 %)
Bottom performing	92	27 (29.35 %)	65 (70.65 %)
Correctly classified		76.56 %	

that these factors contributed little to the predictive ability of the discriminating model across the wide variety of NSD activities. The excluded factors are use of IT, market competition, market differentiation, employee awareness, client's reluctance and inability, policy environment and market acuity.

Table 4.3 presents the discriminant function and the related statistics, which indicate that a good discriminant function had been developed.

The ability of the discriminating and predictive function to classify top performing NSD activities and bottom ones accurately is determined by the classification matrix presented in Table 4.4. The function correctly classified 79.17 % of the NSD samples. The stepwise function correctly classified 85 (85 %) and misclassified only 15 (15 %) from a group of 100 top performing NSD samples. For the 92 bottom performing NSD samples, the model correctly classified 67 (72.83 %) and misclassified 25 (79.17 %). These classification results are higher than they would have been if classified correctly by chance (50.09 %). Thus, the validity of the function's discriminating ability was supported. The U-method (Dillon and Goldstein 1984), which gives an almost unbiased estimate of the classification rate (Sharma 1996), is employed to assess the validity of the discriminant model. As shown in Table 4.4, minor over predictions for the group of top performing NSD samples is 82.00 % and that of the bottom performing group is 70.62 %, which is still within a reasonable range of the stepwise classification rates. Thus, the validity of its discriminating ability was further supported.

4.5 Discussions

According to the objective of this study, we employ the magnitude of the standardized coefficients to interpret the discriminating power of each factor. The larger the size of a coefficient, the greater contribution made to the discriminating power of top performing NSD activities and bottom performing ones. In term of the ranking of the coefficient from high to low, factors discriminating between top performing NSD activities and bottom performing ones are as follows:

Employee competence has the highest standardized coefficient and is, therefore, found to contribute the most to the discriminating power of the function. Specialized expert knowledge, research and development ability, and problem solving know-how are the real products of knowledge intensive business services (Strambach 1997). These services are often thought to depend on people, and particularly their skills and knowledge, to an unusually high degree (Tether and Hipp 2002). Because the knowledge and skills of an organization's employee have become increasingly important to its performance, competitiveness, and innovation (Tharenou et al. 2007), human resources play a key role in new knowledge intensive business service provision (Tether and Hipp 2002). To successfully develop new knowledge intensive business services, firm must employ creative and skilled employees, and treat training as a high priority.

Innovation-supportive culture has the second highest standardized coefficient, which indicating that in developing new knowledge-intensive services it is important to encourage and support openness, creativeness, "stepping out" beyond the norm and risk taking (Anderson and West 1998; de Brentani 2001, Jassawalla and Sashittal 2002). Menor and Roth (2007) argue that innovation-supportive culture is one of new service development competence dimensions. Creating innovation-supportive culture is propitious to enhancing the new service development performance.

Reputation had the third largest standardized coefficient, meaning that reputation is an important micro factor distinguishing top performing NSD activities from bottom performing ones. Broch and Isaksen (2004) argue "Generally the market for knowledge services is characterized by a high degree of information asymmetry between providers and customers. The provider of a knowledge service has by nature more information about the particular area in question than the potential customer". Thus, it is the firm's reputation, rather than the service itself, that often determines buying decisions (de Brentani 1991). de Brentani (1991) argue that focusing new service development efforts on this longer-term objective of gaining a reputation for innovativeness and quality are a route to NSD success since a superior image in the market can only be developed over time.

The final micro factor that exhibited some influence is formal NSD process, although the degree of influence is minor compared to the previous three micro factors. This result suggests that top performing NSD new services are developed through a formal process. Since firms that conduct more rigorous forms of service development activity are found to be more successful at developing and launching new services (Edgett and Parkinson 1994), the effective development of new services requires formal processes and practices (de Brentani 1989; Fitzsimmons and Fitzsimmons 2000).

Trajectories synergy is one of the new variables used in this study. Dosi (1982) define technological trajectory as the direction of advance within a technological paradigm which defines its own concept of "progress" based on its specific technological and economic trade-offs. The search for new product or process is never a random process on the entire set of notional technological opportunities, but depends on paradigms (Dosi 1982). The results of the discriminant analysis suggest that the fitting with trajectories are critical for top performing NSD activities. The development of new service is also never a random process on the entire set of notional opportunities, but depends on professional trajectories, managerial trajectories, technology trajectories, institutional trajectory and social trajectory.

Of the meso factors, only market attractiveness has a significant standardized discriminant function coefficient. While factors relating to other meso factors do not impact success, the results of the discriminant analysis suggest that market attractiveness is critical for the performance of NSD activities. This means that firms targeting large dollar volume market, high growth market and high margin market would result in top NSD performance. Previous NSD success studies (Edgett 1994; Edgett and Parkinson 1994; Ottenbacher et al. 2006) also link market attractiveness to success and failure in NSD. Our findings are again consistent with previous NSD studies in financial services and hospitality industry.

It should be noted that the factors presented here are discussed not only according to the sequence of the strength of their discriminant function coefficients but also according to their category. Employee competence, innovation-supportive culture, reputation and formal NSD process are the four factors found contribute the most to the discriminating power. This means that these micro factors have highest discriminability between top performing NSD activities and bottom performing ones. Hipp and Grupp (2005) argue that "Internal innovation activities in companies are the major stimulating force of (company) growth and change also in the service sector". The final two smallest yet significant factors that exhibited some influence are trajectories synergy and market attractiveness, although the degree of influence is minor compared to the previous micro factors. This result suggests that the successful management of micro, meso and macro factors can create a synergistic effect that leads to top performing NSD activities. Employee competence, innovation-supportive culture, reputation and formal NSD process are central issues in the development of new knowledge intensive business services. Meso and macro factors, however, can only come into effective after and through the success of the micro factors above.

In our studies, seven factors, which are use of IT, market competition, market differentiation, employee awareness, client's reluctance and inability, policy environment and market acuity, are excluded from the final stepwise solution. Johne and Storey (1998) argue "It is interesting, however, that the degree of competition in the marketplace appears not to be related to performance.

Competition is a fact of life in most service industries". Just as competition, institution barriers and use of IT are also the facts of life in KIBS (Howells 2003; den Hertog 2000), which may be a plausible explanation that demonstrates why the two factors do not relate to performance. Bennett and Cooper (1981) suggest that the adoption of the marketing concept philosophy stifles the development and marketing of original new products, and rather encourages the development of product modifications. For KIBS incremental innovations are of major importance (de Jong and Kemp 2003); hence market acuity has no influence on the NSD performance. In incremental innovation, incremental adjustments must be made continuously to meet client needs (de Jong and Kemp 2003), which implies that there are less new knowledge for clients to learning than in radical innovation and that employee are familiar with new service developed by themselves than in radical innovation. Consequently, client's reluctance and inability and employee awareness have little effect on the NSD performance. In our previous study, firms also hold that the client's reluctance and inability is also less important obstacle of innovation in their firms (Liu 2006). Market differentiation is also a fact of life in the most KIBS. It is most of KIBS are not seeking to compete on price (Tether and Hipp 2002; Liu 2006), and often use knowledge resources to building firm-specific competencies that cannot easily be imitated by competitors that give the explanation that market differentiation contribute little to the predictive ability of the discriminating model.

4.6 Conclusion

Though developing successful new services is critical for many companies (de Brentani 1991), past studies of this phenomenon have been concerned almost exclusively with western countries and is limited to production-intensive services and supplier-dominated service (Edgett and Parkinson 1994; de Jong and Vermeulen 2003; Ottenbacher et al. 2006). The primary goal of this research has been to gain a better understanding of what distinguishes the top performing NSD activities within KIBS in Chinese context, and furthermore provides support for the suggestion that macro factors should be included in the research frame work. Our research goal has been accomplished by examining the macro, meso and micro factors which have been identified as potentially determining NSD performance in previous NSD research. By successfully applying stepwise discriminant analysis methodology, this research outlined six factors that significantly affect new knowledge intensive business service development performance in Chinese context. The results of our studies have implications for scholars specialized in service operation management, for manager who want to development new knowledge intensive service.

Of the 13 factors that describe new service development, six factors in all three categories of factors are found to be related to top performing NSD programs in KIBS. According to Ottenbacher et al's suggestion (2006), managers involved in innovation activities should pay attention to these six factors above. The organization characteristics of KIBS are the first category of factors that distinguish between top performing NSD activities versus bottom performing ones. The firms with qualified employee, innovation-supportive culture and good reputation, particularly when combined with formal NSD process, are strong candidates for top performing NSD activities. Though the survey results indicate that the organization characteristic are of fundamental importance, the two other categories of factors have only a moderate effect on the performance of NSD activities. The new services that fit well with the trajectory and that target attractive market can also distinguish between top performing NSD activities versus bottom performing ones.

To sum up, this research contributes to a richer and more systematic understanding of the determinants of top performing NSD activities. In particular, this study examines the influence of macro, meso and micro factors on top performing NSD activities. The findings demonstrate that different categories of factors are important in influence NSD activities. Our results imply the need for KIBS to strategically leverage on the six key antecedents of top performing NSD activities.

References

Anderson N, West MA (1998) Measuring climate for work group innovation: development and validation of the team climate inventory. J Organ Behav 19(3):235–258

Atuahene-Gima K (1996a) Market orientation and innovation. J Bus Res 35(2):93–103

Atuahene-Gima K (1996b) Differential potency of factors affecting innovation performance in manufacturing and services firms in Australia. J Prod Innov Manag 13(1):35–52

Avlonitis GJ, Papastathopoulou PG, Gounaris SP (2001) An empirically-based typology of product innovativeness for new financial services: success and failure scenarios. J Prod Innov Manag 18(5):324–342

Bagozzi RP (1980) Causal models in marketing. Wiley, New York

Bartlett MS (1954) A note on the multiplying factors for various Chi square approximations. J Roy Stat Soc 16(Series B):296–298

Bennett RC, Cooper RC (1981) The misuse of marketing: an American tragedy. Bus Horiz 26(6):51–61

Broch M, Isaksen A (2004) Knowledge intensive service activities and innovation in the Norwegian software industry. STEP REPORT, 03

Chandler GN, Keller C, Lyon DW (2000) Unraveling the determinants and consequences of an innovation-supportive organizational culture. Entrepreneurship Theory and Practice 25(1): 59–76

Chau PYK, Tam KY (1997) Factors affecting the adoption of open systems: an exploratory study. MIS Q 21(1):1–24

Cooper G, Kleinschmidt EJ (1995) Benchmarking the firm's critical success factors in new product development. J Prod Innov Manag 12(6):374–391

Cooper RG, de Brentani U (1991) New industrial financial services: what distinguishes the winners. J Prod Innov Manag 8(2):75–90

Cooper RG, Easingwood CJ, Edgett S, Kleinschmidt EJ, Storey C (1994) What distinguishes the top performing new products in financial services? J Prod Innov Manag 11(4):281–299

Cowell DW (1988) New service development. J Mark Manag 3(3):313–327 de Brentani U (1989) Success and failure in new industrial services. J Prod Innov Manag 6(4):239–258

de Brentani U (1991) Success factors in new developing new business services. Eur J Mark 25(2):33–59

de Brentani U (2001) Innovative versus incremental new business services: different keys for achieving success. J Prod Innov Manag 18(3):169–187

de Jong JPJ, Kemp R (2003) Determinants of co-workers' innovation behaviour: an investigation into knowledge intensive services. Int J Innov Manag 7(2):189–212

de Jong JPJ, Vermeulen PAM (2003) Organizing successful new service development: a literature review. Manag Decis 41(9):844–858

Den Hertog P (2000) Knowledge-intensive business services as co-producers of innovation. Int J Innov Manag 4(4):491–528

Dillon WR, Goldstein M (1984) Multivariate analysis: methods and applications. Wiley, New York

Dosi G (1982) Technological paradigms and technological trajectories: a suggested interpretation of the determinants and directions of technical change. Res Policy 11(3):147–162

Easingwood C (1986) New product development for service companies. J Prod Innov Manag 3(4):264–275

Edgett S (1994) The traits of successful new service development. J Serv Mark 8(3):40–49

Edgett S, Parkinson S (1994) The development of new financial services: identifying determinants of success and failure. Int J Serv Ind Manag 5(4):24–38

Fitzsimmons JA, Fitzsimmons MJ (2000) New service development—creating memorable experiences. Sage Publications, Thousand Oaks

Gallouj F, Weinstein O (1997) Innovation in services. Res Policy 26(4/5):537–556

Hipp C, Grupp H (2005) Innovation in the service sector: the demand for service-specific innovation measurement concepts and typologies. Res Policy 34(4):517–535

Howells J (2003) Barriers to innovation and technology transfer in services. Tech Monitor 20(3):29–35

Jassawalla AR, Sashittal HC (2002) Cultures that support product innovation processes. Acad Manag Exec 16(3):42–53

Johne A, Storey C (1998) New service development: a review of the literature and annotated bibliography. Eur J Mark 32(3/4):184–251

Kaiser H (1974) An index of factorial simplicity. Psychometrika 39(1):31–36

Kaplan RS, Norton DP (1996) Translating strategy into action: the balanced scorecard. Harvard Business School Press, Boston

Kaplan RS, Norton DP (1992) The balanced scorecard: measures that drive performance. Harv Bus Rev 70(1):71–79

Klecka WR (1980) Discriminant analysis. Sage, Newbury Park

Kuusisto J, Viljamaa A (2004) Knowledge-intensive business services and Co-production of knowledge: the role of public sector? Frontiers of e-Business research 2004. In: Conference proceedings of eBRC 2004. Tampere University of Technology and University of Tampere, Finland pp 282–298

Lee K, Shim S, Jeong B, Hwang J (2003) Knowledge intensive service activities (KISA) in Korea's Innovation System. OECD Report, 2

Liu S (2006) The Study on mechanism of knowledge intensive services in knowledge system. Post-doctoral Research Report, Zhongnan University of Economics and Law (in Chinese)

Liu S (2008) The characteristics and evolution of services sector: evidence from P. R. China. Int J Serv, Econ Manag 1(2):183–195

Lovelock C, Gummesson E (2004) Whither services marketing? In search of a new paradigm and fresh perspectives. J Serv Res 7(1):20–41

Lucas BA, Ferrell OC (2000) The effect of market orientation on product innovation. J Acad Mark Sci 28(2):239–247

Lundvall B, Borrás S (2005) Science, technology and innovation policy. In: Fagerberg J, Mowery D, Nelson R (eds) The Oxford handbook of innovation. Oxford University Press, New York

MacMillan IC, McCaffery ML, Wijk GV (1985) Competitors response to easily imitated new products-exploring commercial banking product introductions. Strateg Manag J 6(1):75–85

Menor LJ, Roth AV (2007) New service development competence in retail banking: construct development and measurement validation. J Oper Manag 25(4):825–846

Menor LJ, Roth AV, Mason CH (2001) Agility in retail banking: a numerical taxonomy of strategic service groups. Manufact Serv Oper Manag 3(4):273–292

Miles I, Kastrinos N, Flanagan K, Bilderbeek R, Hertog P, Huntink W, Bouman M (1995) Knowledge intensive business services: users, carriers and sources of innovation, Rappon pour DG13 SPRINT-EIMS

Muller E, Zenker A (2001) Business services as actors of knowledge transformation: the role of KIBS in regional and national innovation systems. Res Policy 30(9):1501–1516

Nunnaly J (1978) Psychometric theory. McGraw-Hill, New York

OECD-EUROSTAT (1997) The measurement of scientific and technological activities. Proposed guidelines for collecting and interpreting technological data. Oslo Manual OECD, Paris

Ottenbacher M, Gnoth J, Jones P (2006) Identifying determinants of success in development of new high-contact services: insights from the hospitality industry. Int J Serv Indus Manag 17(4):344–363

Peaks H, Riihela N (2004) An exploration of inter-functional integration in the new service development process. Serv Ind J 24(6):37–63

Porter ME (1980) Competitive strategy: techniques for analysing industries and competitors. Free Press, New York

Preissl B (1998) Barriers to innovation in services. SI4S Project synthesis work package 2

Sharma S (1996) Applied multivariate techniques. Wiley, New York

Stevens E, Dimitriadis S (2005) Managing the new service development process: towards a systemic model. Eur J Mark 39(1/2):175–198

Storey C, Kelly D (2001a) Measuring the performance of new service development activities. Serv Ind J 21(2):71–90

Storey C, Kelly D (2001b) Measuring the performance of new service development activities. Serv Ind J 21(2):71–90

Storey C, Easingwood CJ (1998) The augmented service offering: a conceptualization and study of its impact on new service success. J Prod Innov Manag 15(4):335–351

Storey CD, Easingwood CJ (1996) Determinants of new product Performance: astudy in the financial services sector. Int J Serv Indus Manag 7(1):32–55

Strambach S (1997) Knowledge-intensive services and innovation in Germany. Report for TSER project, University of Stuttgart

Sunbo J (1997) Management of innovation in services. Serv Ind J 17(3):432–455

Sundbo J, Gallouj F (1998) Innovation in services, SI4S Project synthesis Work package 3/4

Tether BS, Hipp C (2002) Knowledge intensive, technical and other services: patterns of competitiveness and innovation compared. Technol Anal Strateg Manag 14(2):163–182

Tharenou P, Saks AM, Moore C (2007) A review and critique of research on training and organizational-level outcomes. Hum Resour Manag Rev 17(3):251–273

Tsai CT, Chang PL, Chou TC, Cheng YP (2005) An integration framework of innovation assessment for the knowledge-intensive service industry. Int J Serv Technol Manag 30(1/2): 85–104

Van de Ven A (1986) Central problems in the management of innovation. Manag Sci 32(5):590–607

Chapter 5
Knowledge Intensive Service Activities in Chinese Software Industry

Knowledge Intensive Business Services (KIBS) are more and more important for the national economy and have become an important part of national innovation system. This change is called knowledge economy era, postindustrial era or service economy era (Lee et al. 2003). It is important nodes and media of innovation systems, and characters as knowledge intensive inputs and outputs. KIBS are also the most potential growing part in service industry. Software industry is not only a main innovating component of KIBS but also a main innovative source of traditional service industry and manufacture.

Software industry is the main support of innovation in service industry. The importance of innovation in promoting service has been verified by statistics data (OECD 2001). Innovation is the guarantee for the development and restructure of traditional service. As a main part of KIBS, software industry is the base of other services and has changed the technical innovation pattern. The appliance of software technology has changed the traditional operating pattern of not only common service such as logistics, transportation storage but also KIBS such as management consulting, designing and R&D. Software technology broadens the operating extent of time and space, and promotes the rapid development of service.

Software industry is an important technical base of traditional industry informationization. Traditional industry informationization is the process to apply IT to realize information control in designing, producing, product management and operation in order to utilize information resources and improve management, development ability and operating level (Lin and Guo 2002). The basis of this process is to incorporate software with traditional design, manufacture, product and management. In one word, software industry is an important part and base for informationization technology system for traditional industry.

S. Liu, *Innovation Management in Knowledge Intensive Business Services in China*, SpringerBriefs in Business, DOI: 10.1007/978-3-642-34676-7_5,

5.1 The Innovative Source of Manufacturing Informationization

Reconstructing traditional industry with ICT is the process of applying ICT to realize product informationization, process informationization and management informationization, so as to explore and utilize information resources of enterprise and improve management, development ability and operating level (Lin and Guo 2002). Since it is an important task to realize manufacturing informationization, this section analyzes the characteristics of innovative sources of Chinese manufacturing informationization.

5.1.1 Classification of Manufacturing Informationization Technologies

Technology innovation is the process to transform knowledge, skill and matter into satisfying products and services (Fu 1998). Technology innovation can be divided into product innovation and process innovation (Xu 2002). Product innovation is that a new or a prominently improved product enters the market. Process innovation is that a new or a prominently improved process used in business production, which include the change or innovation in equipment or production organization or both. Manufacturing informationization includes product informationization, process informationization and management informationization. The essence of manufacturing informationization is to realize product innovation, process innovation and management innovation on the basis of ICT.

Product informationization is product innovation, which extends the information function of product by processing the inner and external information of the product (Lin and Guo 2002). Process informationization is process innovation, which improves the design of product and production process with the basis of ICT. Management informationization is management innovation, which integrates information flow and business flow together, and then changes the organization structure at the same time. With the realization of the management informationization, the enterprise changes the pyramid management organization into network organization, and decrease the middle level and promoting management efficiency.

We study the relationship between product informationization, process informationization and management informationization by costume manufacturing informationization on the next section.

5.1.2 The Relationship of Different Manufacturing Informationization Technologies

The following steps compose the informationization production process of costume manufacturing. Firstly we should collect information from the management information system to use CAD to design costume, and use CAD system to design pattern; Secondly the cutting CAM cuts the cloth according instructions coming from CAD system; Thirdly the moving CAM send the patterns to the suitable workbench to sew. Fourthly the pressing and finishing CAM press patterns sewed according instructions coming from computer controlling the pressing and finishing device. Finally MIS deals with sale orders and sends the costume to dealers through the logistics channel, and then check inventory to see if we should stock material.

We know that the core of manufacturing informationization is the process informationization through the example of manufacturing costume; the support of informationization is management informationization. Without the process informationization, enterprise can neither satisfy the aim of more effective and higher quality production, nor react the external market change quickly. The purpose of product informationization of lower industry is to realize the process informationization of upper industry.

Since there is bigger gap from the infrastructure of technology and science, technical capacity, education and culture for the developing countries, the self-innovation capacity is based on the technology absorbed and assimilation from developed countries, and technology innovation originates from process innovation, and then comes the product innovation (Xu 2002). So it is more important for Chinese manufacturing to realize informationization in production process.

To sum up, the core of manufacturing informationization is the process informationization for the Chinese manufacturing; the support of informationization is management informationization; the object of informationization is product informationization. The core of informationization in traditional manufacturing is process informationization with help of CAD, CAE, CAM, FMS and NC, etc.

5.1.3 The Characteristics of Manufacturing Informationization

The theory of technology system is represented on the study of factory automation, the theory emphasize the interaction between suppliers, users, universities and research institutes, and suggest that we should reconstruct traditional manufacturing with systematic perspective and identify ICT in a global scope (Carlsson et al. 2002). So we divide the innovative sources of informationization technologies according regions and institutions, the regions are divided into Native province, the other province in China and foreign countries, and the institutions are divided into native enterprise, the other enterprise and public institution. Using data from 84 manufacturing enterprise in China, we study the innovation source of informationization technologies of manufacturing.

With the analysis of 84 manufacturing enterprise informationization, we divide 148 ICT according providing regions and the classification of manufacturing informationization technologies. The 148 ICT divided according region are shown in Table 5.1.

Table 5.1 The 148 ICT divided according regions

	Native province	The other provinces	Foreign countries
Product innovation	3	0	3
Process innovation	9	18	36
Management innovation	18	53	11

Table 5.2 The result of Pearson Chi-Square Tests according regions

χ^2value	P value	Significant level	Degree of freedom	Critical value
44.188	0.000000	0.050	4.000	9.488

Table 5.3 The ratio actual number to theoretical number

	Native province	The other provinces	Foreign countries
Product innovation	4.9333	0.0000	0.0000
Process innovation	0.7048	0.5956	1.7994
Management innovation	1.0829	1.3473	0.4224

Table 5.1 is a 3 × 3 contingency table, we can use Pearson Chi-Square Tests to study the characteristics of manufacturing informationization technology according regions. The results of Pearson Chi-Square Tests are show in Table 5.2.

As shown in Table 5.2, χ^2 value is bigger than the critical value, and the significant level is near zero, so there are significant differences in regional innovative sources of manufacturing informationization technologies.

Now we divide actual number of manufacturing informationization technology of each classifications and regions by its theoretical number, the results are shown in Table 5.3.

As shown in Tables 5.1 and 5.3, the main regional source of product information technology innovation is the native provinces; enterprises tend to use the native province product information technology. The main regional source of process information technology innovation is the foreign countries; enterprises tend to use the foreign countries process information technology. The main regional source of management information technology innovation is the native countries; enterprises don't tend to use the foreign countries management information technology.

With the analysis of 84 manufacturing enterprises information technology, we divide 148 ICT according providing institutions and the classification of ICT in manufacturing information technology. The 148 ICT divided according institutions are shown in Table 5.4.

Table 5.4 is a 3 × 3 contingency table, we can use Pearson Chi-Square Tests to study the characteristics of manufacturing information technology according institution. The results of Pearson Chi-Square Tests are show on Table 5.5.

Table 5.4 The 148 ICT divided according institution

	Native enterprise	The other enterprise	Public institutions
Product innovation	3	0	0
Process innovation	7	49	7
Management innovation	9	68	5

Table 5.5 The result of Pearson chi-square tests according institutions

χ^2 value	P value	Significant level	Degree of freedom	Critical value
22.0143	0.0002	0.0500	4.0000	9.4877

Table 5.6 The ratio actual number to theoretical number

	Native enterprise	The other enterprise	Public institutions
Product innovation	7.7895	0.0000	0.0000
Process innovation	0.8655	0.9839	1.3704
Management innovation	0.8549	1.0490	0.7520

As shown in Table 5.5, χ^2 value is bigger than the critical value, and the significant level is near zero, so there are significant differences in institutional innovative source of manufacturing information technology.

Now we divide actual number of manufacturing information technology of each classifications and institution by its theoretical number, the results are shown in Table 5.6.

As shown in Tables 5.4 and 5.6, the main institutional source of product information technology innovation is the native enterprises; enterprises tend to research and develop product information technology by itself. The main institutional source of process information technology innovation is the other enterprises, but enterprises tend to use the institutions as the sources of process information technology. The main institutional source of management information technology innovation is the other enterprises; enterprises also like to use the management information technology developed by the other enterprises.

5.1.4 Conclusion

Manufacturing process informationization includes mechanization technology, automatization technology and informationization technology, and is comprehensive result of research, design and development about the technologies. The research and development of informationization production equipment has the characteristics of high technology, high cost, high investment, high-risk, high-specialty and the professional knowledge that make general enterprise flinched. Since the business of informationization production equipment in China has poor technology capacity, we have to rely on the foreign manufacturing suppliers in ICT adopted in production process.

Software technology is the core of management informationization, and which can be widely used, so we have a comparative technical advantage in domestic software industry. The big enterprises can also development new management software by themselves. This is the main reason that management information- ization technology mainly relies on domestic corporations or the own enterprises.

Product informationization is the core technology that cannot be outsourced, so enterprise mainly relies itself to research and develop the product informationization technology.

5.2 Performance of Software Industry in China

Software industry is an important node and media of innovation systems, and characters as knowledge intensive inputs and outputs. It is important to study the effect of inputs of software industry and analyze the bottleneck affecting the performance of software industry.

5.2.1 The Indictors Measuring the Inputs and Outputs of Software Industry

Software industry is KIBS with high R&D inputs and good innovation perfor- mance (Lee et al. 2003). In order to well describe the development level of software industry, we should integrate innovation indicators with economic indi- cators in different areas. According to our study of innovation systems (Liu 2003), we start with inputs and outputs in innovation and choose R&D personnel, R&D fund and the number of the invention patent granted to measure the input and development capability in software industry, and then we measure the gross benefit of software industry with total profit in all areas. By this means can we take software industry as a decision-making unit (DMU) with two-input and two-output units. Here the input is the number of R&D personnel and fund respectively, the outputs are profit and the number of inventions patent granted.

As discussion above, the software industry of each city can be seen as a DMU which is shown as Fig. 5.1.

So we can use the method of DEA to study the performance of software industry for each city in China.

Fig. 5.1 Inputs and outputs of software industry

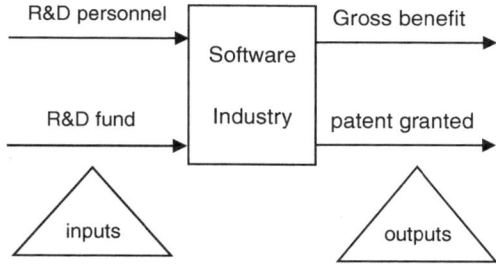

5.2.2 The Theory of DEA

Data Envelopment Analysis (DEA), developed from the concept of "relative efficiency estimate" which is introduced by Charnes and Cooper, is a systematic analysis method to evaluate the relative efficiency of a set of DMUs of the same type (Charnes and Cooper 1978). C^2R Model of DEA can evaluate the scale effectiveness and technical effectiveness at the same time. In another word, an effective DMU in C^2R Model of DEA is both of adequate scale and high technology level (Wei 1988). DEA is a non-parametric technique for measuring and evaluating the relative effciencies of a set of entities, called DMUs, with the common inputs and outputs. Examples include school, hospital, library and, more recently, whole economic and society systems, in which outputs and inputs are always multiple in character. The DEA model classifies the units into two groups efficient and inefficient.

Assume there are m_1 inputs and m_2 outputs in the n DMUs of the same type. We mark input and output vector with $x = (x_1, x_2, \ldots, x_{m1})$ and $y = (y_1, y_2, \ldots, y_{m2})$ respectively.

We set the DMU j_0 input and output vector to x_{j0} and y_{j0} respectively, and then we can evaluate the relative effectiveness of DMU j_0 in the following C^2R Model:

$$\text{Min} \quad [\theta - \varepsilon(e^t s^- + e^t s^+)]$$

$$s.t. \quad \sum_{j=0}^{n} \lambda_j x_j + s^- = \theta x_{j0}$$

$$\sum_{j=0}^{n} \lambda_j y_j - s^+ = y_{j0} \tag{5.1}$$

$$\lambda_j \geq 0, j = 1, \ldots, n$$

$$s^- \geq 0, s^+ \geq 0$$

where ε is the Archimedes infinitesimal number, θ^* is the optimal solution.

If $\theta^* = 1$, DMU_{j0} is weak DEA effective;

If $\theta^* = 1$ and $s^{*-} = 0$, $s^{*+} = 0$, DMU_{j0} is DEA effective.

Table 5.7 The scores of indicators in each city

Cities	R&D personnel	R&D fund	Gross benefit	Invention patent granted	Cities	R&D personnel	R&D fund	Gross benefit	Invention patent granted
Beijing	100.00	100.00	89.21	90.99	Jinan	8.11	6.56	15.65	34.23
Tianjin	11.80	6.77	8.45	42.34	Qingdao	5.03	7.80	1.57	23.42
Shijiazhuang	1.70	1.07	0.63	7.21	Zhenzhou	1.42	0.46	0.83	7.21
Taiyuan	0.39	0.01	0.25	1.80	Wuhan	14.86	10.62	18.52	29.73
Shenyang	3.30	1.97	0.69	17.12	Changsha	7.56	4.04	10.86	21.62
Dalian	2.17	0.81	4.28	3.60	Guangzhou	23.08	28.47	27.04	1.80
Changchun	3.85	3.03	14.38	20.72	Shenzhen	30.25	36.58	42.52	87.39
Harerbin	9.49	10.72	20.72	2.70	Chongqin	4.48	1.83	6.18	22.52
Shanghai	23.70	36.19	72.21	100.00	Chengdu	19.92	9.81	100.00	22.52
Nanjing	15.24	10.03	96.93	76.58	Guiyang	0.57	0.32	0.03	3.60
Hangzhou	1.25	1.11	23.74	2.70	Kunming	2.97	0.76	8.01	36.04
Fuzhou	7.75	6.75	4.20	15.32	Xian	7.95	6.65	8.31	13.51
Xiamen	1.80	1.93	1.30	2.70	Lanzhou	5.52	1.86	2.44	8.11
Nanchang	5.80	4.14	11.54	10.81					

Table 5.8 The performance and the shadow price of inputs in each city

City	Optimal value of C^2R	Shadow price		City	Optimal value of C^2R	Shadow price	
		R&D personnel	R&D budget			R&D personnel	R&D fund
Taiyuan	1.00	21.57	179.83	Jinan	0.43	1.23	0.00
Hangzhou	1.00	7.64	0.42	Tianjin	0.41	0.85	0.00
Kunming	1.00	1.13	8.78	Qingdao	0.40	1.99	0.00
Nanjing	1.00	0.66	0.00	Dalian	0.30	0.89	9.91
Shanghai	1.00	0.42	0.00	Changsha	0.28	1.32	0.00
Beijing	1.00	0.10	0.00	Xiamen	0.27	5.55	0.00
Chengdu	1.00	0.01	1.00	Nanchang	0.24	1.72	0.00
Guiyang	0.92	17.44	0.00	Wuhan	0.23	0.67	0.00
Shenzhen	0.63	0.33	0.00	Fuzhou	0.19	1.29	0.00
Zhengzhou	0.56	7.05	0.00	Xian	0.18	1.26	0.00
Changchun	0.56	2.60	0.00	Lanzhou	0.16	1.81	0.00
Shijiazhuang	0.47	5.90	0.00	Harerbin	0.12	1.05	0.00
Shenyang	0.47	3.03	0.00	Guangzhou	0.08	0.43	0.00
Chongqing	0.45	2.23	0.00				

As C^2R Model is a linear programming model, the economic meaning of input constraining shadow price is that the number of the increment of the target functional value is induced by one-unit increase in input. This increment is DEA effective. Then we can access the scarcity of input relative to DMU. The higher is shadow price; the more effectiveness can the added input enhance DMU. This indicates that this input contributes a lot to the target value and this input is a scare resource.

5.2.3 Data Processing Method and Result

The data of the R&D personnel, R&D fund, the invention patent granted and the total profit is from Statistics Bureau of the People's Republic of China.

In order to suit the computing and analysis convenience, the indictors are standardized to the scores from one to one hundred. We define zbi as the statistic value of the indicator for city i ($i = 1, 2, \ldots, 27$), $zbfsi$ is the score of this indicator in the observed year for city i. The score can be computed as following:

$$zbfs_i = \frac{100 \times zb_i}{\underset{i}{\mathrm{Max}}(zb_i)} \qquad i = 1, 2, \ldots, 27 \qquad (5.2)$$

According to the formula (5.2), the scores of indicators in each city is shown in Table 5.7.

We use C^2R Model to measure the performance in the software production of different cities. So the optimal value of the model is the relative productive efficiency in the city j_0, the shadow price of one input denotes how much can this input effect the productive efficiency of software industry. The results are shown in Table 5.8.

According to the analysis of s^- and s^+ in the result of DEA, in the seven weak DEA effective cities (the optimal value of the C^2R Model is 1) above, only Taiyuan, Hangzhou and Kunming are DEA effective, the three cities satisfy $\theta^* = 1$, $s^{*-} = 0$, $s^{*+} = 0$.

According to the analysis of C^2R Model and the theories of DEA, We can analyze the performance and the influential factors of software industry in each city.

From Table 5.8 and the result analysis, we can see that performance of software industry is the best in the city of Taiyuan, Hangzhou and Kunming. R&D personnel and R&D fund have a great contribution to the innovation and profit capability, and the two inputs are fully utilized.

Nanjing, Shanghai, Beijing and Chengdu are weak DEA effectiveness. This reveals that the performance of software industry in the three cities is also high. However, they have a gap relative to Taiyuan, Hangzhou and Kunming. In these cities except Chengdu, the R&D personnel contributes more than that of fund.

Though Guiyang is not weak DEA effective, it has a comparative higher performance; R&D personnel also contributes more than that of capital.

The other cities have a relative low productive efficiency; R&D personnel contribute less than that of fund.

5.2.4 Conclusion

Services are often thought to depend on people, and particularly their skills and knowledge, to an unusually high degree (Tether and Hipp 2002). Specialised expert knowledge, research and development ability, and problemsolving know-how are the real products of knowledge-intensive services (Strambach 1997).

Human resources are the carrier of knowledge, especially the tactic knowledge, as knowledge is hold by the specific person. Human resource is the key factor for the development of KIBS since KIBS mainly depends on the input of knowledge (Miles et al. 1995). Software industry is an important component of KIBS, the development of which also depends on high-level professionals. Then technical professionals are the key factor of software industry and they are also the main bottleneck of the development of software industry till now. So we should strengthen the cultivation and fetch of software professionals. This measure is the key of developing Chinese software industry.

Based upon the above findings, we developed some policy implications. Firstly, governments should provide training programs that enable service workers to work better with innovations. Along with human resource development programs, it is recommended that training courses involving service management and innovation management in services should be provided so that service suppliers and their users can improve their absorption capacity, which may lead to building innovation capability in the long run. Secondly, a comprehensive education policy, emphasising multidisciplinary and lifelong learning, is key to developing such capital. Policies should focus on developing the abilities of individuals to communicate, adapt to change, solve problems, network and interface effectively with clients and colleagues. Finally, It is important to improve the social milieu, since it is argued here that knowledge is fundamentally centered on the individual, social relations and context will be of great significance in the creation and dissemination of knowledge (Howells and Roberts, 2000).

References

Carlsson B, Jacobsson S, Holmen M, Rickne A (2002) Innovation systems: analytical and methodological issues. Res Policy 31(2):233–245

Charnes A, Cooper WW, Rhodes E (1978) Measuring the efficiency of decision making units. European J Oper Res 12(6):429–444

Fu J (1998) Technology innovation. Tsing Hua university press, Beijing

Howells J, Roberts J (2000) From innovation systems to knowledge systems. Prometheus 18(1):17–31

Lee K, Shim S, Jeong B, Hwang J (2003) Knowledge intensive service activities (KISA) in Korea's innovation system, OECD Report, 2

Lin J, Guo C (2002) The key of traditional industry reforming by information technology—product informatization. Chinese Mech Eng 9:769–771 (in Chinese)

Liu S (2003) The Study on Measure and Comparison of Regional Innovation System. Bei-Hang University, The thesis for the Doctorate (in Chinese)

OECD (2001) Innovation and productivity in services. OECD Publications, Paris

Strambach S (1997) Knowledge-intensive services and innovation in Germany, Report for TSER project, University of Stuttgart

Tether BS, Hipp C (2002) Knowledge intensive, technical and other services: patterns of competitiveness and innovation compared. Technol Anal Strateg Manag 14(2):163–182

Wei Q (1998) Relative efficiency estimate DEA methode—new field of operational research. China Renmin University Press, Beijing (in Chinese)

Xu Q (2002) Management of research development and technology-based innovation. Higher education press, Beijing (in Chinese)

Miles I, Kastrinos N, Flanagan K, Bilderbeek R, Hertog P, Huntink W, Bouman M (1995) Knowledge intensive business services: users, carriers and sources of innovation, Rappon pour DG13 SPRINT-EIMS, March

Index

B
Balanced scorecard, 17, 20, 37, 50

C
Characteristic, v, vi, 1, 2, 4, 7, 8, 10, 13–15, 24, 43, 46, 47, 59, 64–67

D
Data envelopment analysis, 68, 69, 71, 72
Determinants, vi, 24, 43–45, 54, 59

I
Innovative characteristics, 2, 4, 7
Innovative model, 9, 11
Innovative performance, 32, 39

K
Knowledge intensive business service, v, vi, 1–4, 7, 8, 10, 15, 23–25, 27, 30, 44, 46, 50, 56, 57, 63

M
Manufacturing informationization, 64–66

N
New service development, v, vi, 1, 19, 26, 29, 32, 34, 36, 43–48, 51, 52, 58
New service development activities, vi, 52

S
Service innovation, 8–15, 43, 44, 46, 47, 52
Software industry, 8, 63, 68, 69, 71, 72

T
Top performance, 44, 49–51, 54, 55–59
Typology, 3

S. Liu, *Innovation Management in Knowledge Intensive Business Services in China*, SpringerBriefs in Business, DOI: 10.1007/978-3-642-34676-7,